Computing and Systems Technology Division 2017

Core Programming Area at the 2017 AIChE Spring Meeting and 13th Global Congress on Process Safety

San Antonio, Texas, USA
26 – 30 March 2017

ISBN: 978-1-5108-4145-1

Printed from e-media with permission by:

Curran Associates, Inc.
57 Morehouse Lane
Red Hook, NY 12571

Some format issues inherent in the e-media version may also appear in this print version.

Printed by Curran Associates, Inc. (2017)

For permission requests, please contact AIChE
at the address below.

AIChE
120 Wall Street, FL 23
New York, NY 10005-4020

Phone: (800) 242-4363
Fax: (203) 775-5177

www.aiche.org

Additional copies of this publication are available from:

Curran Associates, Inc.
57 Morehouse Lane
Red Hook, NY 12571 USA
Phone: 845-758-0400
Fax: 845-758-2633
Email: curran@proceedings.com
Web: www.proceedings.com

TABLE OF CONTENTS

480365 Constrained Grey-Box Multi-Objective Optimization Framework for Optimal Design of Energy Systems

Abstract:

Energy systems are characterized by large, diverse number of components in which they form an integrated complex multi-scale network. The complexity of these multi-scale models is further amplified with the generation of large number of data, which makes them harder to express with mechanistic formulations. In such models, derivative information is often unavailable or unreliable, and the direct use of optimization methods is usually rather prohibitive. [1] Thus, the global optimization of complex energy systems from an economic and sustainability perspective poses a formidable challenge.

Derivative-Free Optimization (DFO) methods are commonly utilized for the optimization of models that lack the closed-form equations or models that strongly rely on input-output data. We have previously introduced the constrained grey-box optimization algorithm called ARGONAUT [2] that couples tractable surrogate approximations, which accurately represent any unknown correlations, with the *state-of-the art* Mixed-Integer Nonlinear Programming (MINLP) global optimization solver ANTIGONE. [3] In this work, we further expand the existing algorithm to handle mixed-integer programming and multi-objective optimization problems, and test the proposed framework on a case study based on the energy system design for commercial buildings such as a supermarket. [4] We provide solutions to two cases; (a) optimal design based on the single-objective economic behavior or the environmental impact (b) optimal design based on the multi-objective design criteria, simultaneous optimization of economic and environmental behavior. We demonstrate that our framework enables optimization of expensive simulation-based models under multiple competing objectives in a computationally efficient way. The results are presented in the form of Pareto-frontier, compare favorably to the model-based solution in [4].

References:

[1] Boukouvala, F.; Misener, R.; Floudas, C. A., Global optimization advances in Mixed-Integer Nonlinear Programming, MINLP, and Constrained Derivative-Free Optimization, CDFO. *European Journal of Operational Research* **2016,** *252*, (3), 701-727.

[2] Boukouvala, F.; Floudas, C. A., ARGONAUT: AlgoRithms for Global Optimization of coNstrAined grey-box compUTational problems. *Optimization Letters* **2016**, 1-19.

[3] Misener, R.; Floudas, C., ANTIGONE: Algorithms for coNTinuous / Integer Global Optimization of Nonlinear Equations. *Journal of Global Optimization* **2014,** *59*, (2-3), 503-526.

[4] Liu, P.; Pistikopoulos, E.N.; Li, Z., An energy systems engineering approach to the optimal design of energy systems in commercial buildings, *Energy Policy* **2010**, *38*, (8), 4224–4231.

480919 Decision-Making Tool for the Optimal Planning and Scheduling of Utility Assets

Abstract:

With the cost of energy driven by the market in many places and the number of environmental regulations increasing, companies are more and more focused on improving their energy management tools to achieve a more efficient allocation of their energy assets while reducing the environmental impact. When to buy and store fuel, when to produce and sell electric power to the grid while satisfying internal demand, when to start-up or shut-down the cogeneration units, are a few of the many questions energy schedulers often face on a daily basis and whose answers have a great impact on the profitability of the site. Tackling these questions successfully is not trivial since it requires not only considering the system in the current state but also future potential states.

With this in mind, in this paper we introduce our Energy Multi-Period Optimizer technology [1], an energy management software (VM-EMPO) for optimal planning and scheduling of utilities assets. VM-EMPO allows scheduling engineers and operators to find the optimal operation of the utility plants on a time horizon when dealing with time dependent constraints. These constraints are often related to fuel tank farms, thermal energy storages, minimum/maximum start/stop time of equipment, and planned out-of-service periods. VM-EMPO offers a robust modeling and optimization engine and an easy-to-use graphical user interface.

We illustrate the benefits of using VM-EMPO in two energy management applications. Namely, the optimal operation of a thermal energy storage and chillers in a district cooling system and the optimal fuel storage management and cogeneration unit operation in the steam system of a petrochemical plant.

[1] Ruiz, J. P., Santollani, O., Optimal Planning and Scheduling of Utility Assets, Industrial Energy Technology Conference (IETC) Proceedings (2016).

479842 Control and Diagnosis of Battery State of Charge

Abstract:

Control and Diagnosis of Battery State of Charge

Resmi Suresh M P[1] and *Raghunathan Rengaswamy[1]

[1]Department of Chemical Engineering, IIT Madras, Chennai, 600036, India

Corresponding author: raghur@iitm.ac.in

In the present world, life without gadgets is unimaginable and almost all of these gadgets use batteries. On the other hand, there are also reports of batteries catching fire or exploding and the reasons for these can be attributed to poor health of the battery due to either internal short circuit or excessive heat (especially due to overcharging) [1,2]. An efficient miniature online diagnostic tool, an accurate model and a controller could help avoid these mishaps by ensuring safe operation of batteries [3]. Battery health can be monitored using a diagnostic tool and this information could be used to update the model of the battery. A controller based on this updated model could be used to control the charging current and prevent overcharging. A framework for online diagnosis and control of batteries as shown in Figure 1 is presented in this work.

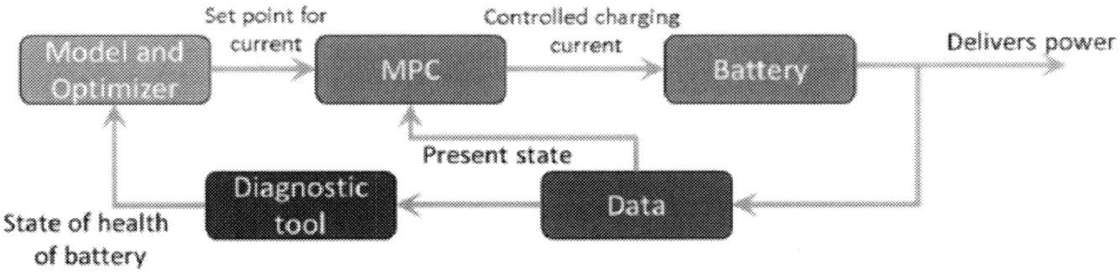

Figure 1: Framework for online diagnosis and control

A diagnostic tool based on impedance information derived using chirp voltage and current data is described in this work [4]. As a first step of battery diagnosis, the chirp technique is applied on a battery model instead of a real battery to understand the mapping between the impedance profiles obtained and battery state of health. A simple model incorporating all the possible failure modes and side reactions which can be solved in real time is necessary for this. The present work uses a model developed based on reaction engineering perspective to characterize parasitic reactions and the electrochemical reactions [5].

An optimization framework to reduce the losses due to side reactions is also proposed in this work. This aims at optimizing the charging current such that reduction in maximum attainable capacity is minimized and charging is achieved within the maximum time allowed. This optimized current is used as a set point for a model predictive controller that can control the charging current at desired levels.

References:

[1] Triggs R. Why phones explode sometimes, and what you can do to protect yourself 2016. http://www.androidauthority.com/what-makes-smartphones-explode-714380/.

[2] Hamill J. Samsung to recall all Galaxy Note 7s after smartphones exploded while charging 2016. https://www.thesun.co.uk/news/1715098/samsung-halts-sale-of-galaxy-s7-after-users-claim-their-smartphones-exploded-whilst-charging/.

[3] Suresh R, Kumar Tanneru H, Rengaswamy R. Modeling of rechargeable batteries. Curr Opin Chem Eng 2016;13:63–74. doi:10.1016/j.coche.2016.08.005.

[4] Bullecks B. Rapid fault detection and mitigation strategies in the low temperature polymer electrolyte membrane fuel cell. Texas Tech University, 2013.

[5] Suresh R, Rengaswamy R. Modeling Failure Modes in Li-Ion Battery. 2015 AIChE Annu. Meet., 2015.

USING AN OPERATIONAL EXCELLENCE PLATFORM TO REDUCE ENERGY AND CARBON LOSS

By Kumar Prashant[+], Steve Hall[+], Miguel Angel Castro[0], Mohammed Zahrani[0], Hani Al Saed[0]
[+]Process Systems Enterprise, [0]Saudi Aramco
Contact: Steve Hall, Tel: +44 7534 721862, Email: s.hall@psenterprise.com

Introduction

Operational Excellence (OE) focusses on sustainable improvement across a business. This large subject area comprises several important elements such as tools, operating procedures and management systems. This paper describes how the authors have sought to implement Operational Excellence. It describes the motivations behind the initiative and shows the pragmatic approach taken. The development of specific tools to meet the project objectives are described, together with the performance targeting methodology adopted. Data visualisation is shown to be critical and the development of dashboarding elements shows how data are pushed to the various stakeholders in a pro-active way. Some case studies are presented - OE platforms for fuel gas management and heat exchanger fouling management.

Manufacturing Landscape

The work described in this paper has been applied in the Middle East region. This region is highly dependent on natural gas as its raw material, yet low oil and rising gas prices are stifling the region's growth. Global competition is increasing and costs are higher. There are also a glut of new projects coming on stream soon[1], so the commercial pressures have never been greater. As a result, Middle East operators are looking for any way to improve operations and address falls in bottom-line profits.

Another growing issue is the slow-down in innovation projects. Generally speaking, all the consultants and EPC companies tend to use the same software tools, technology has become a commodity and competing operators work with the same engineering companies. Carving out a unique commercial strategy which puts them clearly ahead of their competitors is therefore more difficult.

When looking to gain a commercial advantage, the preference will always be to focus on operational improvements where little/no investment is required and risks are low. This is consistent with the region's historical operations too, where previously, depressed energy pricing has allowed assets to be operated non-optimally, leaving room now for such operational improvements. When considered alongside that fact that tools to model and identify improvements in operating plants have significantly improved over recent years, together with data visualisation technologies, there are certainly opportunities there to improve. The main challenge remains – how to harness these various technologies, making them work together in a seamless way, to convey the results to the user in simple terms and to drive informed improvement decisions.

What is Operational Excellence

Operational excellence is an all-encompassing approach to identifying improvements now, actioning change to realise those improvements and then critically, to regularly ensure that such

[1] Chemical & Engineering News, Volume 94 Issue 2 | pp. 9-15, January 11, 2016

improvements are managed. Too often, optimisation projects are implemented and within 6 months, either forgotten about through staff changes/other operating foci, or in the case of real-time optimisation systems, just switched off.

Figure 1. Operational excellence

The clear focus of OE is an inclusive approach to achieve sustainable improvements. In its truest form, OE covers a wide range of areas but even today, despite all the initiatives in this area, there is still no single tool which addresses all satisfactorily.

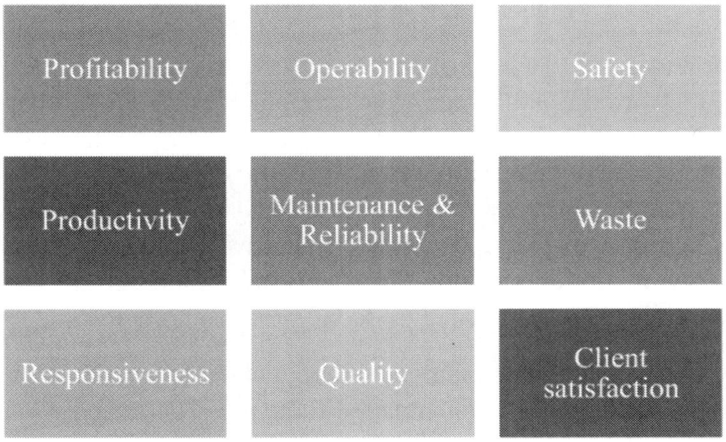

Figure 2. Elements of an integrated OE system

Beyond the capabilities of the available tools, there remain some key challenges, the largest one being company culture. OE approaches are successful when everyone sees plant profitability as their problem, where staff are proactive in sorting issues out and where the contribution of the front-end staff and senior managers are both considered equally. Of course, culture can be formalised into management systems and operating procedures but the culture must go beyond such documented approaches.

A key part of harnessing the contributions of individuals and teams to operational improvements is that they need to have access to information which is appropriate to their level of expertise and presented in an appropriate way i.e. the right metrics in the right format.

OE approaches must be able to adapt quickly to changes in operating constraints, economic climate and product demand and the insights of staff across the business can most often deliver advice on the adjustments required to maximise performance and deliver a return on capital across all assets.

Implementation Strategy

Some work has been done[2] in the area of capturing the knowledge and experience of individuals in driving process improvements. However, there are still gaps between the integrated implementation of process knowledge management and some of the more easily implemented data/modelling tools. For the current initiatives, emphasis was placed on integrating available tools and supplementing their effectiveness, and involving the people element as an extra rather than as a fully integrated part of the solution. This was also deemed to be the quickest and easiest way to achieve some benefit.

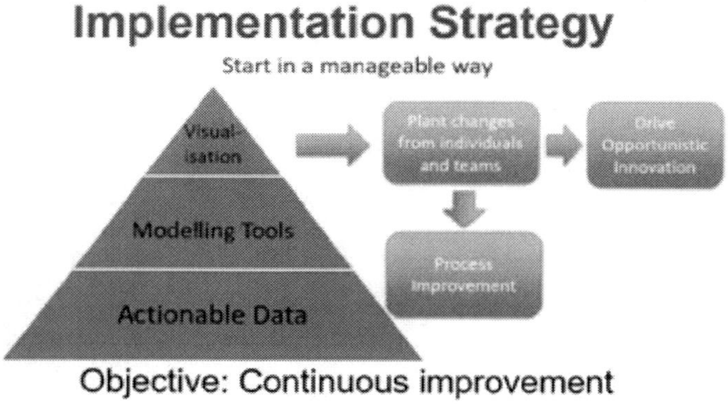

Figure 3. OE implementation strategy

The above Figure shows the approach taken here. It represents a fairly standard approach to process improvement. Data is retrieved from the plant systems, either DCS or historian. The data is fed into modelling tools which simulate current conditions and then determine the best operating conditions. The results are then presented to the user in a way which facilitates action – identifying what exactly is needed to improve operation. The results can be reviewed by individuals and teams, which then recommend changes to improve operations. Such changes will likely be operational only – to vary flows, temperatures, heating duties, etc. However, the visualisation may also spur additional innovations, perhaps to make structural changes to the process through some sort of revamp project.

Even though the approach in Figure 3 is fairly standard, the way the various steps are implemented is open to variation and the authors have implemented some novel approaches to make the best use of Operational Excellence tools, as highlighted in Figure 4.

[2] Use process knowledge management systems to accelerate innovation, Steve Hall, and Phil Mahoney, Hydrocarbon Processing, October 2016

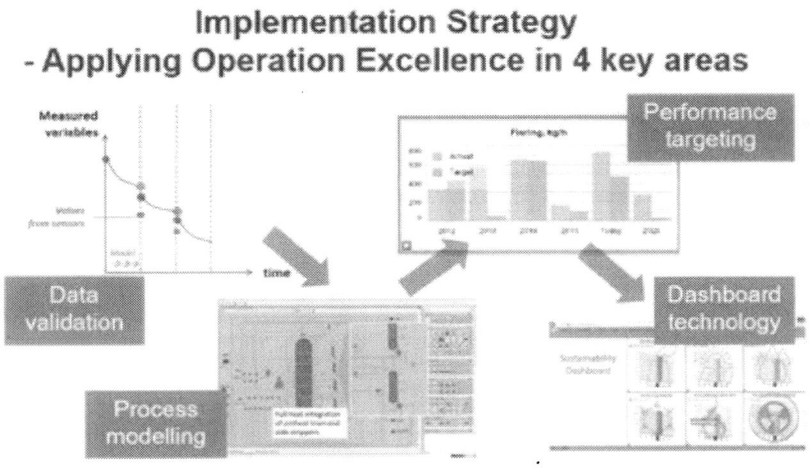

Figure 4. Excellence in the implementation of tools

The various elements of this approach are described in the following section.

Using Tools to Meet Our Needs

Tools form a key tranche of any OE system. However, there is much potential flexibility in how the tools are implemented both in terms of what they do and how they are integrated together. The aim in this work is to not only produce short-term gain, but also to keep the benefits sustainable, to continually ensure the plant is kept operating at its optimum point.

The following 4 areas represent the new technology/approaches being used.

Excellence in Data Validation

A major challenge in using plant data is how to ensure any process model is consistent and accurately reflects the process systems. Instrumentation is notoriously unreliable and so we use techniques and approaches to define our model. The data validation and reconciliation techniques chosen can vary depending on the specific situation and the extent and type of data available. The approach can be a simple averaging of readings over time or something more complicated. PSE has developed State Estimation techniques which assign a probability distribution for measured variables. For example, confidence levels are likely high for temperature readings (although offset values are common) and low for flow measurements, depending on the sensor type. Another complicating area for data reconciliation is that the underlying process system may be changing, perhaps due to feed variation or control action. This presents an additional challenge to reconcile data within a dynamic process system. PSE has developed a linear quadratic equation approach[3] to address this, accounting for the statistical noise inherent in measured variables over time, and implemented this using an Extended Kalman-type filter.

Another unique aspect of the current approach is that the data and modelling system learns as it goes. In the case of Fouling Management, where advanced equations are used to predict heat exchanger fouling rates into the future, the trajectory of fouling rate is regularly determined and re-calibrated

[3] First-principles modelling for optimization of process operations, Costas Pantelides, Foundations of Computer Aided Process Operations / Chemical Process Control, Tucson, Arizona, January 8-12, 2017

using current plant data and the various descriptive equations for fouling are regularly re-tuned to maintain accuracy. The process of state estimation is represented in Figure 5.

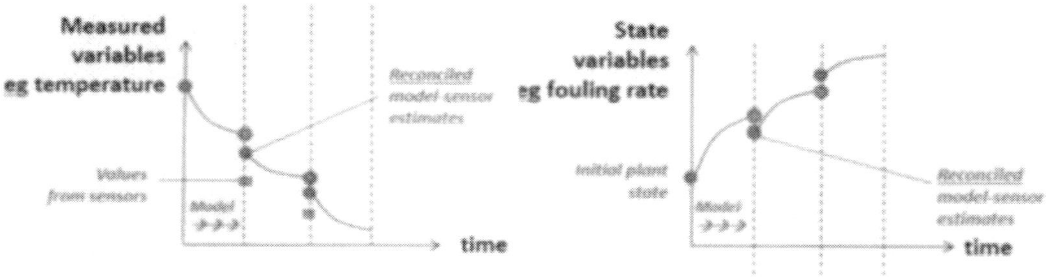

Figure 5. State estimation techniques applied to tune heat exchanger fouling rates

Excellence in Process Modelling

The data validation techniques described earlier require a process model to be used in the validation process. The current projects use PSE's gPROMS ProcessBuilder platform. This is an equation-oriented process modelling environment[4] which has some significant advantages over the more traditional sequential modular approaches. Firstly, the equation-orientated approach allows data reconciliation and validation to be easily achieved. Secondly, the equation-orientated model provides a consistent modelling environment and a uniform model both for steady-state modelling and later on for optimisation, simply by adding in an objective function such as finding the minimum cost operating conditions subject to certain operating constraints.

gPROMS Process Builder is particularly useful in the case studies described later. For example, in the Fouling Management deployment, the modelling environment is used to optimise not only at a single point in time (i.e. what are the best operating conditions today), but also optimise over the production run, perhaps in a typical refinery, 5 years until the next shutdown. The equation-oriented approach allows optimisation to determine what heat exchangers need to be cleaned and when, subject to several constraints such as scheduled cleaning periods, maximum number of cleans allowed, guaranteed minimum production rates, etc. Such optimisations are very difficult to achieve in sequential-modular tools.

Excellence in Performance targeting

The focus of this part of the OE implementation is to first establish current performance, establish the best possible performance based on operating constraints and highlight the gaps. Also to highlight how to get from the current operating state to the optimum state. We use a principle of 'avoidable' and 'inevitable' losses and this is quite a useful concept when thinking about carbon or energy loss. It can be easily visualised for plant staff, both in terms of amount and also costs.

One major challenge in this area is how to most effectively advise operations staff on what to do to improve operation. We take a 2 tier approach – the first is to recommend based on continuous variables – to keep everything operating at the same on-off state but vary the loads to achieve the improvement. The second level is to recommend based on integer optimisation – to allow units to be switched on and off. In our case studies, the economics of fuel gas systems is very much dependent

---

[4] Equation-oriented process modelling technology: recent advances & current perspectives, Constantinos C. Pantelides, Maarten Nauta and Mark Matzopoulos, 5th Annual TRC- Idemitsu Workshop "The role of collaboration in enhancing Refining Industry R&D", Sas Al Nakhl, Abu Dhabi, UAE, February 11-12, 2015

on both levels, whereas in the case of Fouling Management, one generally optimises on the first tier of only continuous variable optimisation.

Another critical area is in constraints. Many optimisation systems produce results and operator advice which are either impractical or give too many advice elements leading to confusion and uncertainty. In the current work, much effort has been placed in rationalising the improvement actions to keep the number to a minimum – only focussing in on the big improvement items, and giving clear directions.

Excellence in Dashboard Technologies

We all can express our satisfaction or dissatisfaction with our experience using software tools. The way we interact and the intuitive nature of the tools we use can turn us on or off to what they are saying. To design a tool which conveys critical information in a pleasing way, to both operators and senior managers, is a significant challenge in itself and one which we spend much time developing. Companies are even now using gaming techniques such as virtual reality[5] to enhance the user experience and train operators. The other problem we have is the shear amount of data that is now available to us. How to use such vast data sources is another issue[6] beyond this current presentation.

The case studies described here use the latest in Dashboard technologies, both to display live plant data, show problem areas such as where excess flaring is occurring or heat exchanger fouling will soon become an issue, and show the improvement actions required either now or in future. The key elements of dashboards are the ability to limit the information seen based on user login details, the ability to drill-down to more details so the information is only ever delivered at an appropriate level required by the user. Most critically though, the dashboard allows results to be viewed in a web browser, providing enterprise-scale deployment of the information. In terms of OE, this is a significant requirement.

For the current projects, extensive consultation has taken place, with staff from engineering, operations and maintenance involved in specifying the content of the dashboards and report generators. Behind the scenes, the database structure is also designed carefully to ensure the right data is stored for the end purpose. The role of Dashboard databases in this context is not to reproduce the information within data historian, but to limit data to the key improvement objectives. It also means that some calculated results may indeed be fed back to the data historian, with unique new Tag IDs, for use elsewhere in plant analysis tools.

Case Study: Fuel Gas System

In the first case study, the challenge is to identify what needs to be done to minimise flaring on a petrochemical plant and so reduce carbon loss. The various steps outlined above are followed. In this case, there are higher level dashboards which are accessed by senior managers and planners which display information on various sustainability objectives, for example intensity of use of water, waste, energy and atmospheric emissions. However, the dashboard which the operator sees is limited to flaring on the site as this is the operator's specific focus area. The dashboard is presented here:

---

[5] The application of virtual reality to chemical engineering and education, John Bell and Scott Fogler, AIChE Meeting, Miami, Florida, November 16-20, 1998

[6] Smart plant applications in crude distillation units, Steve Hall and Kumar Prashant, AIChE Meeting, San Antonio, Texas, March 26-30, 2017

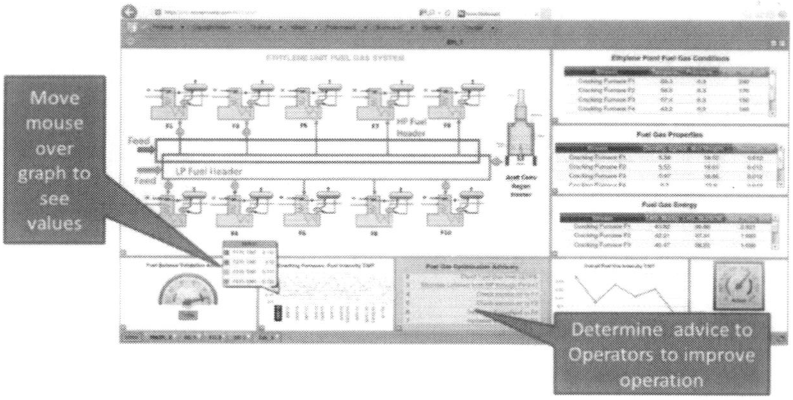

Figure 6. Dashboard for fuel gas systems to minimise flaring and reduce carbon loss

Case Study: Heat Exchanger Network Fouling Management

Our second case study focuses on the management of fouling in heat exchanger networks. The four key steps outlined above were followed.

In **data validation**, parameter estimation techniques are used to reconcile live plant data. The calculation of fouling rates and trajectory of fouling in each heat exchanger is designed to be regularly recalculated using current plant data. This ensures that the latest plant data is incorporated into the fouling calculations, maximising accuracy which later feed into the correct actions being taken to maximise ongoing profitability.

For **process modelling**, PSE's gPROMS ProcessBuilder platform is used which is a true equation-oriented system. This allows the data validation and reconciliation step to be seamlessly achieved alongside handling the complex equations required to determine which heat exchangers need to be cleaned and when, to be efficiently calculated over a given time horizon. It also allows optimisation algorithms to be included to improve overall operation, changing operating parameters to address both fouling and energy consumption.

For **performance targeting**, gPROMS ProcessBuilder provides the ability to rapidly calculate optimal operating conditions for each day in the whole operating window, providing a very detailed economic assessment to be made and informed operating decisions presented to the user. The penalty of operating each heat exchanger is evaluated. Also, the penalty of operating the crude unit in a certain way, for example with specific pump-around flows or stream splits, is assessed, allowing such information and corrective actions to be documented in the associated operating database and presented to the user.

For **dashboard technologies**, the Fouling Manager system is deployed in 2 forms. The first is an off-line tool for engineering staff to conduct what-if studies and for planners to assess maintenance schedules. There is also an on-line tool which presents the current state-of-health of heat exchangers and of the crude unit itself and advises what, if anything, needs to be done either today or at some point in the future to maintain maximum profitability and keep atmospheric emissions under control. Both versions of the tool use the same, live plant data which has been validated and reconciled.

Dashboards for the on and offline tools are shown below:

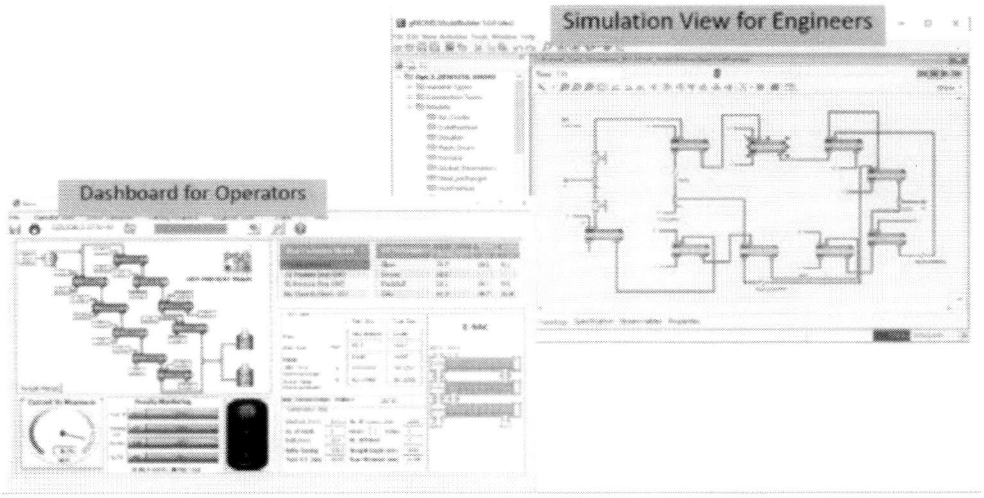

Figure 7. Dashboards for the heat exchanger fouling management tool

Future Developments

The current implementation of the OE platform presents useful data to the users and informs the user as to what actions will improve operations and profitability. Further work in this area needs to focus on capturing the ideas and insights from all stakeholders themselves[2]. There are often practical issues associated with heat exchanger network operation, for example in the area of reliability or other features not easily incorporated into a simulation model. The intent is to extend the OE platform capabilities to capture such knowledge. Another example, in the case of heat exchanger fouling, is to learn from the cleaning regimes in each heat exchanger, to assess what cleaning methods work and develop a consistency of cleaning approach with procedures.

Summary and Conclusions

This paper has described an approach to implementing OE into a working platform. Although the steps taken to create the platform are typical of industry approach, four specific areas of excellence have been developed, namely data validation and reconciliation techniques using state estimation methods; process simulation methods using equation-orientated approaches for easy convergence of complex optimisation problems across operating days and time; targeting techniques which recognise inevitable and avoidable losses of carbon and/or energy; dashboard technologies which give the user a positive user experience and provide the appropriate information at the right level for all stakeholders, from manager to operator. By implementing these specific approaches into an Operational Excellence platform, the tools and technology is integrated to give maximum benefit to reduce both carbon losses through reduced flaring and reduced energy consumption through better management of fouling in heat exchanger networks.

Impact of Computational Mesh on CFD Combustion Predictions

Ryan Sperry
Undergraduate Research Assistant (ryanasperry@gmail.com)

Bradley Adams
Associate Professor (brad.adams@byu.edu)

Department of Mechanical Engineering
Brigham Young University

Introduction

Use of computational fluid dynamics (CFD) for combustion modeling has become widespread throughout the petrochemical industry. Commercial CFD software such as CD-Adapco's StarCCM+ and ANSYS's Fluent are routinely used by designers, vendors, consultants and end users to evaluate burner and furnace design and performance. For example, CFD models have been used to predict flame length, flame impingement, heat flux profiles, heat transfer, gas properties, and pollutant concentrations [1-6]. Because combustion is a complex interaction of turbulent mixing, chemical reactions and heat transfer, development of accurate CFD combustion models requires familiarity with both combustion principles and modeling options in the CFD software. Specifically, modelers are required to select appropriate computational grids, boundary conditions, turbulent flow and chemical reaction sub-models, and numerical solution techniques. The specification of grids and boundary conditions and selection of appropriate sub-models is complicated by the numerous options available in commercial CFD codes. Some of these options include structured or unstructured grids and pyramid, tetrahedron, hexahedron or polyhedron cell types. These options can aid modelers in accurately representing the variety of physical conditions that occur in different combustion processes, but to inexperienced users the preponderance of gridding and sub-model options can be confusing, particularly when specification of an incorrect grid or boundary condition or selection of an incorrect sub-model can impact the accuracy of the overall prediction. These risks of incorrect model setup or sub-model selection can be mitigated as users gain additional training, experience, and exposure to CFD modeling.

The objective of this study was to evaluate the effects of different mesh types and mesh refinement options on the results of CFD combustion modeling. The CFD software used in the study was CD-Adapco's StarCCM+. The study involved five different meshes of a single burner in a furnace. Fuel type, stoichiometry and firing rates were not varied in this study. The only independent variable manipulated was the meshing conditions, with all other boundary conditions and inputs staying fixed across all cases. Results illustrate the impact of the three key elements of mesh quality – skewness (the amount of deviation from a normalized equilateral angle in a computational cell), smoothness (the rate of change in cell size), and aspect ratio (the ratio of cell longest edge to shortest edge length). The metrics used to compare the cases were: boundary radiation heat flux profile, total heat transfer, thermal efficiency, gas exit temperature, and qualitative comparisons of flame shape and mixing patterns.

Study results are targeted to new or inexperienced CFD modelers and are expected to provide some general guidelines for effective meshing that will help novice practitioners avoid common mistakes made by new users.

Case Description

The application selected for study was a single vertical cylindrical (VC) heater burner in a test furnace. The test furnace was 3.2004 m (10.5 ft) wide by 3.6576 m (12 ft) deep by 12.6492 m (41.5 ft) tall, with four 16.8275 cm (6.625-in) diameter vertical (floor to roof) cooling tubes spaced evenly along the back wall; tube centerlines were 0.4572 m (1.5 ft) from the back wall. The furnace exit was a 0.9144 m by 0.9144 m (3 ft x 3 ft) opening centered over the burner. For simplicity, all furnace walls were assumed to be adiabatic. Cooling tubes were assigned a fixed temperature of 310.9 K (100 °F) and emissivity of 0.85. A 2.931 MW (10 MBtu/hr) burner was centered in the box. The burner was representative of low NOx burners used in VC heaters, with ten staged firing tips arranged symmetrically outside a sloped burner tile at an effective diameter of 72.39 cm (28.5 in). Air flowed upwards in the center of the burner through a 42.55-cm (16.75-in) diameter throat at a mass flow rate of 1.1688 kg/s (9276.7 lb/hr). Figure 1 shows the base outline of the furnace geometry. Note that the sketch is not drawn to scale, but has key dimensions. Fuel was 100% methane at 60 °F distributed evenly through the ten fuel tips entering at a total mass flow rate of 0.05879 kg/s (466.57 lb/hr) for all fuel tips. Inlet air temperature was 288.71 K (60 °F). The burner was operated at a stoichiometry of 1.15, with 3% excess O_2 (dry). There was no premixing of fuel and air in the burner.

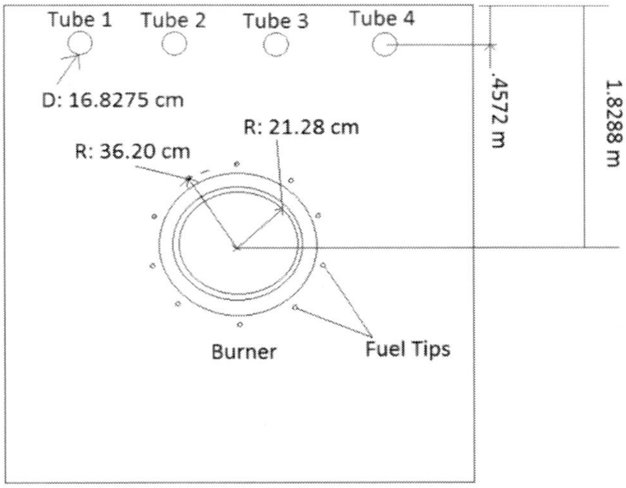

Figure 1. Burner and furnace dimensions.

All cases had the same burner, cooling tube and furnace geometry, the same boundary conditions, and the same fuel and air properties. The following standard StarCCM+ sub-models were used for all cases: k-epsilon turbulence sub-model, Presumed Probability Density Function (PPDF) chemical reaction sub-model with equilibrium chemistry, discrete ordinates radiation solver with a weighted sum of gray gas sub-model for the participating medium.

Modeling Approach

Five computational meshes were developed in order to compare the effects of number of grid cells (1 million vs 2 million cells), cell type (structured trimmed-cell vs polyhedral), and grid refinement location (mixing zone refinement vs inlet zone or structure geometry refinement). A summary of the meshes for the five cases is given in Table 1.

Table 1. Summary of Computational Meshes

	Case 1	Case 2	Case 3	Case 4	Case 5
Number of cells	1 million	2 million	1 million	2 million	1 million
Cell type	Structured trimmed-cells	Structured trimmed-cells	Polyhedral cells	Polyhedral cells	Structured trimmed-cells
Refinement focus	Mixing zone	Mixing zone	Mixing zone	Mixing zone	Geometric shapes

Figure 2 shows the computational mesh with the different cell types and levels of refinement for Cases 1-5. Cases 1, 2 and 5 have a structured mesh with a predominantly hexahedral cell type ("trimmed cell"), whereas Cases 3 and 4 have an unstructured mesh with polyhedral cell type. Both mesh types utilize local mesh refinement. The figures also illustrate the different levels of skewness, smoothness, and aspect ratios (defined earlier) for the meshes. Cases 1-4 focus mesh refinement in regions of high shear flow and mixing, i.e., near the fuel jet inlets and above the burner tile. Transition in cell size is relatively smooth, cell skewness is generally low, and aspect ratios in these high-shear zones are moderate. Conversely, Case 5 focuses mesh refinement near the flow-structure boundaries, i.e., the cooling tubes, furnace walls and floor, and burner. This results in fewer cells in the fuel jet inlet and burner mixing zones and increased skewness and aspect ratios in the mesh as well as significantly decreased smoothness in cell size transitions. This type of mesh might result from too much focus on resolving surface geometries and too little focus on flow mixing and reaction zones internal to the computational domain.

Case 1 Case 2 Case 3 Case 4 Case 5

Figure 2. Side-by-side comparison of computational meshes.

As noted above, all cases had similar inputs and boundary conditions. All five cases were run out to 5000 iterations. This number of iterations was chosen to achieve relatively similar convergence between the cases in order to provide a relatively fair comparison of results.

Results and Discussion

As mentioned above, each of the five case results were compared by the following metrics: total heat transfer to the process or cooling tubes, thermal efficiency, exit temperature, boundary radiation heat flux profiles, flame shape, and flow patterns. Table 2 shows the data for gas exit temperature, heat transfer, thermal efficiency, and approximate flame length for each of the cases.

First to be analyzed is the total heat transfer and exit temperatures. Table 2 shows there was a general trend followed among all cases that as the total heat transfer increased, the average exit gas temperature decreased. This is consistent with the case boundary conditions since all surfaces besides the exhaust outlet and the tubes are adiabatic making the only paths for energy to exit being these two sources. As more energy from the furnace is transferred from the gas to the tubes, then the gas should have less energy and thus be cooler as it passes through the outlet. Thermal efficiency was calculated as the total heat transfer divided by the firing rate. The firing rate was kept constant across all cases so that the efficiency increased and decreased in direct correlation to the total heat transfer.

Table 2. Summary of Predicted Results.

	Case 1	Case 2	Case 3	Case 4	Case 5
Mesh Type	Structured	Structured	Polyhedral	Polyhedral	Poor refinement
Targeted Cell Count	1,000,000	2,000,000	1,000,000	2,000,000	1,000,000
Actual Cell Count	982,263	2,097,728	1,081,727	1,970,394	1,070,531
Average Exit Gas Temperature (K)	1571.1	1556.0	1584.7	1561.8	1550.9
Total Heat Transfer to Cooling Tubes (MW)	0.9179	0.9495	0.9156	0.9439	0.9625
Firing Rate (MW)	2.9307	2.9307	2.9307	2.9307	2.9307
Thermal Efficiency	31.32%	32.40%	31.24%	32.21%	32.84%
Approximate Flame Length (m)	5	4.75	3.25	5	5

This first analysis shows that Cases 1 and 3 with similar cell count numbers, but different cell shapes, have similar results whereas Cases 2 and 4 with double the cell count have different results from Cases 1 and 3, but similar to each other. Cases 2 and 4, like Cases 1 and 3, have similar cell count, but different cell shape. Case 5 is the outlier with notably higher total heat transfer. This discovery could lead to an assumption that Cases 1 and 3 were unable to reach as accurate of solution due to the lower number of cells and that the cell shape doesn't have as much of an effect on the results as do cell count and refinement locations, but further investigation of additional results is needed to verify this assumption.

This leads to further investigation of the heat transfer and more specifically the heat flux profiles at the cooling tubes. Figure 3 is a graph of the boundary radiation heat flux profile on the cooling tubes for each of the cases with respect to the distance along the tubes (y-direction). Cases 1-4 have similar flux profiles that have lower flux magnitude near the floor and increase with elevation as the gas temperatures from the flame increase. It can be seen that Case 5 yet again was an outlier with a heat flux profile that was generally flatter and higher across the entire tube. This would logically translate into an unreasonably higher total heat transfer as seen in Table 2.

To further investigate the cause of these heat flux profiles, flame shape was addressed as a possibility. For each of the cases, an estimate of flame shape was generated by creating an iso-surface at any point in the grid where there was a concentration of 1500 ppm of CO. This flame shape was overlaid with a scalar function of temperature to show were the flame is hotter and colder. Figure 4 below is a side-by-side comparison of each of the flame shapes.

First off, Case 2 and Case 4 looked fairly similar. This followed the trend of similar heat transfer and similar heat flux profiles. Case 4 appeared to be slightly cooler than Case 2 at the base of the flame on the side closes to the tubes. This correlates with the smaller heat flux in Case 4 on the lower part of the tubes. Other than that, the rest of the flame height and temperature appeared to be very close giving a possible reason for the difference in thermal efficiency of the two cases being only 0.2%.

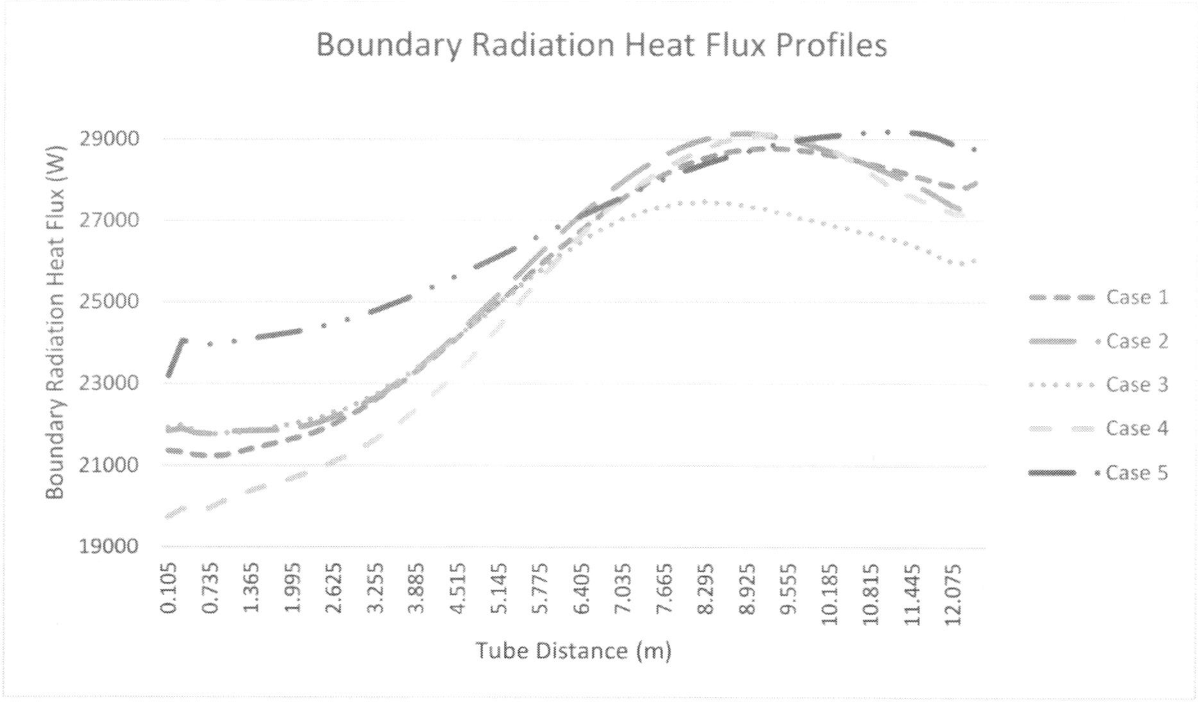

Figure 3. Cooling tube boundary radiation heat flux profiles.

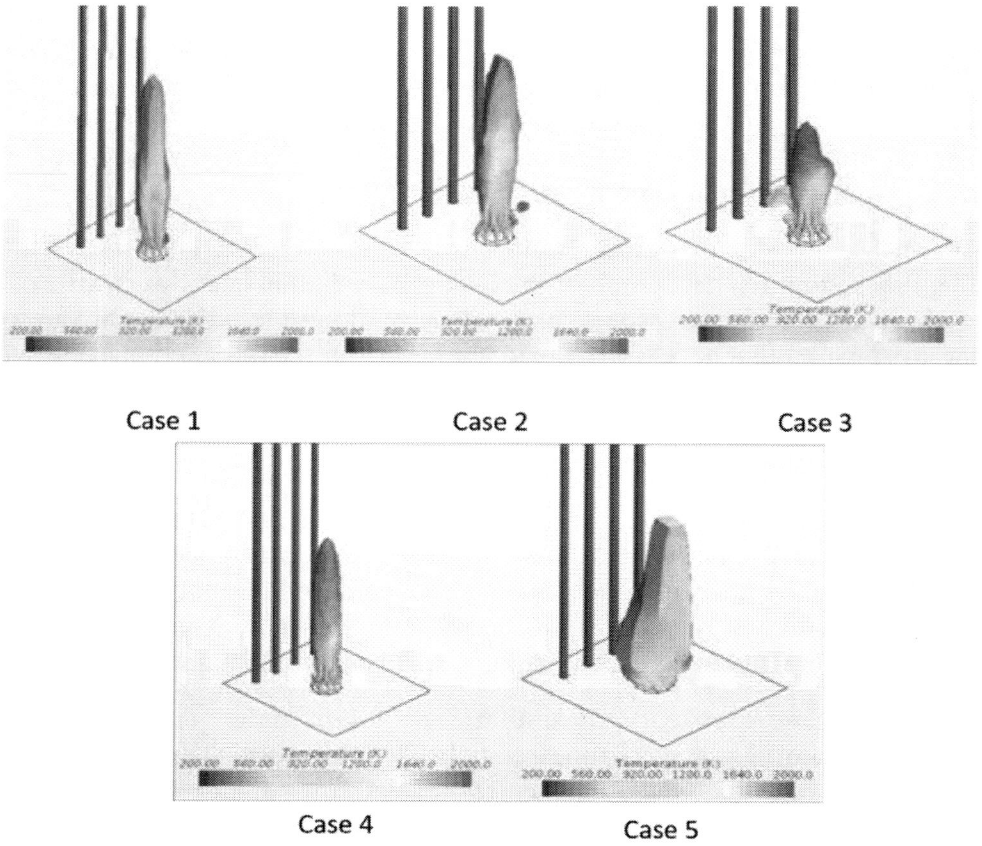

Figure 4. Flame shapes based on iso-surface of 1500 ppm of CO.

Cases 1 and 3 told a slightly different story. Case 1 had a much longer and thinner flame shape than Case 3. This is a plausible cause of the differing heat flux profiles as seen in Figure 3. The bases of the flame showed very similar temperatures close to the tubes which would translate directly to a similar heat flux. The longer shape of the flame from Case 1 could result in a higher heat flux peak and the shorter flame from Case 3 could cause a much lower peak which is what is seen in Figure 3. While the flame shapes seem very different, this seemed to have little to no effect on the total heat transfer with only a 0.08% difference between the two cases.

Comparing the flux profiles as a function of refinement levels between Cases 1 and 2 and Cases 3 and 4 show little differences, but a few variations should be noted. Cases 1 and 2 show very little difference with almost identical flame shapes and very similar heat flux profiles with Case 1 just slightly below Case 2. The only notable difference is the total heat transfer which leads to a thermal efficiency for Case 1 at 1.08% lower than that of Case 1. This is consistent with the heat flux profiles. Cases 3 and 4 show a few more differences. The flame shapes are notably different with Case 3 being much shorter. After analyzing Case 3 and seeing that the flame shape has been relatively constant for the final 2000 iterations as well as the residuals, it is concluded that Case 3 was unable to resolve the ideal flame shape with the level of refinement given. This seemed to have a very large effect on the heat flux profile with Case 3 having a lower peak. This, like the comparison of Cases 1 and 2, has been translated into a lower total heat transfer which is apparent with a difference in thermal efficiency of 0.97%

Now to consider Case 5, the outlier. It can be noted immediately that the flame shape from Case 5 was drastically different than that of all the other cases. It was a lot wider especially at the bottom than Cases 1, 2 and 4. Case 5 also has a more uniform temperature over the flame at a medium high level of about 1650 K. This can be translated to a flatter and higher heat flux profile which is what is seen in Figure 3. In addition this would lead to a higher total heat transfer as mentioned before and as seen in Table 2.

From further qualitative analysis of Case 5, it can also be seen that the mixing patterns are very different which could be the cause of this unique flame shape. Figure 5 shows the scalar field of fuel mixture fraction (i.e., ratio of fuel to fuel plus air) right around the burner for Case 1 and Case 5. The scale range shown is from 0 to 0.2 with red representing higher fuel fractions and blue lower fuel fractions. In this case, a value of zero would be 0% fuel or 100% air, and a value of one would be 100% fuel and 0% air. As can be seen in the Case 1 plot, the fuel inlet jets maintain a coherent structure beyond the top of the burner tile. This produces better penetration into the inlet air creating key mixing above the burner. On the other hand, due to the poor mesh refinement above the fuel tips and burner in Case 5, the fuel jets are not preserved and key mixing above the burner is not achieved. These differences in mixing result in the different flame shapes shown in Figure 4 and the different heat flux profiles shown in Figure 3. Burners are designed to operate in a manner shown in the Case 1 plot of Figure 5. The inadequate mesh refinement around the burner in Case 5 results in an unrealistic representation of the fuel and air mixing, and subsequently the flame shape and heat flux profile.

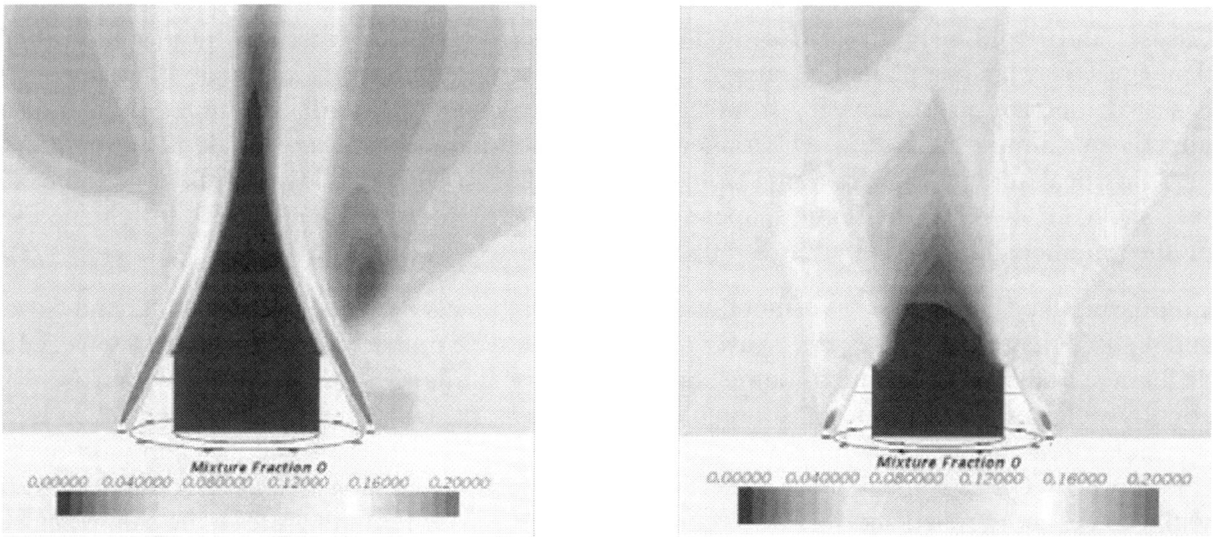

Case 1 Case 5

Figure 4. Predicted fuel mixture fraction for Case 1 and Case 5.

Summary

This study has examined the impact of cell type and mesh refinement on predicted combustion results. Mesh refinement in regions of high velocity gradients was shown to be the most critical factor in achieving realistic simulation results. In Case 5 where mesh refinement was focused near the geometric shapes of the cooling tubes and burner tile, insufficient mesh refinement existed in the fuel-air mixing zones around and above the burner to accurately predict fuel jet penetration and fuel-air mixing, resulting in inaccurate predictions for flame shape and heat flux profiles at the cooling tubes.

Differences in predicted incident flux, total heat transfer and flame shape between the highest levels of mesh refinement (~2 million cells) between the structured trimmed-cell mesh and the unstructured polyhedral mesh were minimal. At lower mesh levels of 1 million cells, there were slightly greater discrepancies between the structured trimmed-cell mesh and the unstructured polyhedral mesh, particularly with respect to the flame shape and upper furnace flux profile predicted in Case 3. Interestingly, the differences in thermal efficiency and total heat transfer were greater between Case 2 and Case 1 (2 million versus 1 million cells) than between Case 2 and Case 5 (2 million versus 1 million cells with poor refinement). This trend was also observed with Case 4 and Case 3. This suggests both cell count and local mesh refinement are important to achieving accurate combustion predictions, but that there is minimal difference between the results with differing mesh types as long as the cell count and refinement are sufficient to represent the expected jet mixing behavior.

These results also suggest that while the flame shape does impact the boundary flux profile as expected, the impact on thermal efficiency may be less well correlated. Thus, total heat transfer results should not be used to assess flame shape behavior; boundary flux profiles provided better correlation to flame shape.

References

1. Adams, B., Cremer, M., Murphey, J., "Use of CFD in Evaluating Pyrolysis Furnace Design," 2015 Spring Meeting & 11th Global Congress on Process Safety, AIChE, ISBN 978-0-8169-1089-2, Paper 2b, April, 2015.
2. Adams, B., Cremer, M., Olver, J., "Impact of High-Emissivity Coatings on Process Furnace Heat Transfer," 2015 Spring Meeting & 11th Global Congress on Process Safety, AIChE, ISBN 978-0-8169-1089-2, Paper 85b, April, 2015.
3. Tang, Q., Denison, M., Adams, B., and Brown, D., "Towards Comprehensive Computational Fluid Dynamics Modeling of Pyrolysis Furnaces With Next Generation Low NOx Burners Using Finite-rate Chemistry," Proceedings of Combustion Institute, Volume 32, Issue 2, Pages 2649-2657, 2009.
4. G.D. Stefanidis, B. Merci, G.J. Heynderickx, G.B. Marin, "CFD simulations of steam cracking furnaces using detailed combustion mechanisms," Computers and Chemical Engineering, 30, 635-649, 2006.
5. Adams, B. and Q. Tang, "Modeling Flame Behavior and Low NOx Emissions in Cracking Furnaces," 15th IFRF Member's Conference, Pisa, Italy, June 13-15, 2007.
6. Brown, D., Ma, J., Adams, B., "Development of an Improved Prediction Model for Chemical Process Furnaces," AIChE 2003 Spring National Meeting, New Orleans, LA, March 31, 2003.

480049 Improving Operational Flexibility and Efficiency in Industrial Steam Systems

Abstract:

Modern industrial sites are often integrated with multiple processes and plants. Often in such a site, there is one or more centralised heat and power (CHP) stations to produce electricity and steam, which get distributed at various pressure levels of the steam network. Processes and plants take steam from the steam grid, and return excess steam to the grid. Operating the steam system can often be challenging due to the configuration complexity and the constraints on the operations. Constraints on the operation can severely restrict the scope for optimization. In extreme cases, the steam system can suffer from a form of 'grid-lock', in which operational flexibility can lead to steam venting. Additionally, there are many practical constraints presented in an integrated industrial site, including flow direction, steam quality assurance at steam pipe end and minimum flow through certain devices. These constraints cause difficulty to adjust operations to optimise for better operational efficiency.

This study uses steam system mathematical modelling and optimisation to improve steam system operational flexibility of the steam system, as well as to reduce energy cost of the steam system. With a site-wide steam network model, it is possible to quantitatively identify operational constraints that inhibit the scope for optimization and in extreme cases the causes of the 'grid-lock' problems. The approach can then identify where flexibility can be increased and suggest project ideas to unlock the network. Rigorous equipment models, such as the steam pipeline hydraulic model, have been integrated in the study to identify challenges of key practical constraints. The proposed operation strategies to reduce energy cost have also been verified with such detailed equipment models.

A case study based on a total site energy study with a large-scale industrial site in Western Europe is presented. Two aspects have been covered in this case study, one is to operationally optimise steam system operation cost, and the other is to tackle certain 'grid-lock' issues in the steam network. Practical aspects such as condensate prevention, handling dynamic demand swing in steam headers, throughput flowrate limits of steam pipelines and assuring steam temperature and pressure for end process users etc. are considered in the study. Thus, a series of operational optimisation strategies and retrofit project ideas to improve steam configuration flexibility are proposed.

481453 A Systematic Approach to Alarm Design with Application to the Tennessee Eastman Problem

Abstract:

Abstract for 2017 AIChE Spring Conference

A systematic approach to alarm design with application to The Tennessee Eastman Problem

JOSHIBA ARIAMUTHU VENKIDASALAPATHY*, COSTAS KRAVARIS and M. SAM MANNAN

Artie McFerrin Department of Chemical Engineering

Texas A&M University

College Station, Texas 77843-3122, USA

***+1(979)402-9823, joshvenkat03@tamu.edu**

When a process plant faces an abnormal situation, the alarm management system that is meant to assist process operators, may hinder effective remedial measures by simultaneously displaying several alarms, making it nearly impossible for the operator to identify the critical variable that needs immediate attention. This condition is known as 'alarm flooding'. This study focuses on efficient alarm identification which is the first critical step in designing the alarm system of a process plant.

We formulate a systematic approach to alarm identification wherein a subset of measured variables is chosen for configuration in the alarm system. The approach may be outlined as (i) Identification of potential hazardous scenarios, (ii) Simulation of these scenarios in order to quantify process measurement response time, (iii) Formulation of a mixed integer linear programming problem (MILP) that is solved using MATLAB. The solution to this problem gives a set of process variables that minimizes the time taken for abnormality detection while taking into account all potential hazards. Additionally, this optimal set ensures that the number of alarms activated for any fault is not more than a threshold number.

The formulation is as follows:

Objective function : $\underset{x}{\text{minimize}}\{\max_{j}\{\min_{i}[\text{Non-zero } T_{ij}]\}\}$

where,

$X = [x_i]$

$$x_i = \begin{cases} 1 \text{ if the measured variable is configured to alarm system} \\ 0 \text{ if the variable is not configured to alarm system} \end{cases}$$

$i = \{1, 2, \ldots\ldots, n\}$, n is the total number of measured variables

$j = \{1, 2, \ldots\ldots, m\}$, m is the number of faults

$T \in R^{n \times m}$ is the Response Time Matrix obtained using simulation,

T_{ij} being the time taken by the ith variable to detect the jth fault.

T_{ij} is assigned as 10000 if the jth fault cannot be detected by i^{th} variable.

such that, $1 \leq \{\text{Number of alarms}\} \leq N$

For this particular study, the MILP routine 'intlinprog' has been used which is based on the branch and bound algorithm.

The proposed approach is then applied to a benchmark industrial plant control problem, the Tennessee Eastman Process Control Problem [1], a well-defined simulator of a chemical process plant which is extensively used in process control research. It consists of a reactor, separator, condenser, stripper and a recycle compressor. The exothermic reactions involved and its operation under pressurized conditions make it a safety-critical plant, and hence a potential application for the proposed algorithm.

A number of major hazards were identified for this plant including control valve 'stiction' and sensor faults. The abnormal scenarios were simulated using the closed loop simulator [2], [3] in a MATLAB environment. The simulations were carried out, one scenario at a time. An important observation based on the results obtained is that the number of alarms for any scenario should be not less than 4, in order to detect any abnormal situation soon enough to provide the operator with adequate response time. The optimal set of process variables identified to be configured in the alarm system are 'Reactor pressure', 'Separator liquid Level', 'Stripper Liquid Level' and 'Purge gas composition- G component'.

References

[1] Downs, James J., and Ernest F. Vogel. "A plant-wide industrial process control problem." Computers & chemical engineering 17, no. 3 (1993): 245-255.

[2] Ricker, N. Lawrence. "Decentralized control of the Tennessee Eastman challenge process." Journal of Process Control 6, no. 4 (1996): 205-221.

[3] Ricker, N. Lawrence. "Tennessee Eastman challenge archive." [Simulink Model], Retrieved from http://depts.washington.edu/control/LARRY/TE/download.html (accessed November 21, 2016).

480348 A Model Based Multiparametric Programming Strategy for the Integration of Design and Control Under Uncertainty

Abstract:

Simultaneous strategies for process and control design are expected to deliver integrated and intensified processes with reduced costs and energy requirements [1]. The problem of simultaneous design and control is recieving an increasing amount of attention in the open literature [2]. This extensive research has yielded different methodologies, but none that utilize a comprehensize framework that begins with a high fidelity model, robust optimization techniques, and a model predictive controller (MPC).

In this study, we present the foundation of a simultaneous strategy that addresses many of the above challanges [3]. In particular, it follows a *(i)* mixed integer dynamic optimization model representation of the integrated process design implemented in a modeling environment (e.g. gPROMS®), *(ii)* the development of a suitable approximate model, amenable for control studies, *(iii)* a multi-parametric optimization and explicit model predictive control step, which allows for the development of explicit expressions of receding horizon control strategies as a function of the design vairables, and *(iv)* a closed loop validation step where the optimal integrated design and control expressions of step *ii* are included within the original high fidelity model in step *i*. To illustrate the technique discussed, we present three case studies of different complexities: a continuously stirred tank reactor (CSTR), a binary distillation column, and a residential cogeneration heat and power (CHP) system.

References

1. Pistikopoulos, E.N., Diangelakis, N.A., Towards the integration of process design, control and scheduling: Are we getting closer? (2015) Computers and Chemical Engineering, 91, pp. 85-92.
2. Aneesh, V., Antony, R., Paramasivan, G., Selvaraju, N., Distillation technology and need of simultaneous design and control: A review (2016) Chemical Engineering and Processing: Process Intensification, 104, pp. 219-242.
3. Pistikopoulos, E.N., Diangelakis, N.A., Oberdieck, R., Papathanasiou, M.M., Nascu, I., Sun, M. PAROC - An integrated framework and software platform for the optimisation and advanced model-based control of process systems (2015) Chemical Engineering Science, 136, pp. 115-138.

480644 Process Safety in Concentrated Solar Plants (CSP)

Abstract:

Heat Transfer Fluid (Synthetic Oil) and Thermal storage fluid TSF (sodium nitrate and potassium nitrate) can result into Process Safety Incidents of level Tier-1 and Tier-2 in Concentrated Solar Plants.

NOMAC is a leading O&M service provider of proven, safe and cost-effective O&M services of desalination and power plants in the Middle East, Africa and Europe having a portfolio of about 20,000 MW of power generation. NOMAC Operates & Maintains combine cycle, simple steam cycle, solar photovoltaic and concentrated solar thermal plants (CSP) and moving towards wind energy projects as well.

Heat Transfer Fluid (Synthetic Oil) and thermal storage fluid TSF (sodium nitrate and potassium nitrate) are being used in Concentrated Solar Plants to exchange heat with water. As a result super-heated steam is generated, which runs steam turbine to produce electricity through generator. Both these substances HTF & TSF have got excellent heat transfer capacity. But at the same time there is a high potential of Fire and Explosion. As TSF can oxide abruptly and can result into explosion and Fire if exposed to Air. Both these chemicals are being used above the threshold quantity defined by API-754. So this can result into TIER-1 and TIER-2 Process Safety Incidents. This paper discusses different incidents related to HTF and TSF in our CSP sites, which can be a lesson learn for exponentially growing CSP market, and can save people, equipment and environment from the Hazardous nature of HTF and TSF.

We observed serious HTF heat exchanger leakages where both these chemicals are coming in contact to exchange heat. This paper will enlist the root causes and contributing causes that jeopardized process safety. What controls we have adopted to get rid of these incidents in future. Incidents belongs to our sites in Morocco and South Africa.

2017 AIChE®

SPRING MEETING & 13TH GLOBAL CONGRESS ON PROCESS SAFETY

March 26–30, 2017
Henry B. Gonzalez
Convention Center
San Antonio, TX

17 AIChE Spring Meeting, San Antonio, TX
13TH GLOBAL CONGRESS ON PROCESS SAFETY

---

17 AIChE Spring Meeting, San Antonio, TX
13TH GLOBAL CONGRESS ON PROCESS SAFETY

CCC

Using Simulation and Control System Emulation

Wayne Jacobson
Luca D'Orazi
Compressor Controls Corporation

2017 AIChE® SPRING MEETING & 13TH GLOBAL CONGRESS ON PROCESS SAFETY

17AIChE Spring Meeting, San Antonio, TX — 13TH GLOBAL CONGRESS ON PROCESS SAFETY

Control System Design Goals

- Reliable Machinery and Process
  - Prevent unnecessary process trips and downtime
  - Minimize effect and duration of process disturbances
  - Prevent surge, overspeed, overheating, and possible associated damage
  - Automate startup and shutdown process
- Efficient Machinery & Process Operation
  - Maximize process throughput
  - Minimize setpoint deviation
  - Optimize loadsharing of multiple units
  - Minimize antisurge recycle or blow-off
  - Operate at lowest possible energy levels

17AIChE Spring Meeting, San Antonio, TX — 13TH GLOBAL CONGRESS ON PROCESS SAFETY

Traditional Simulation for Turbomachinery

- Simulation has been used for verification of equipment
  - Startup, shutdown & trip scenarios
  - Driver power
  - Recycle valve capacity & requirement for hot gas bypass

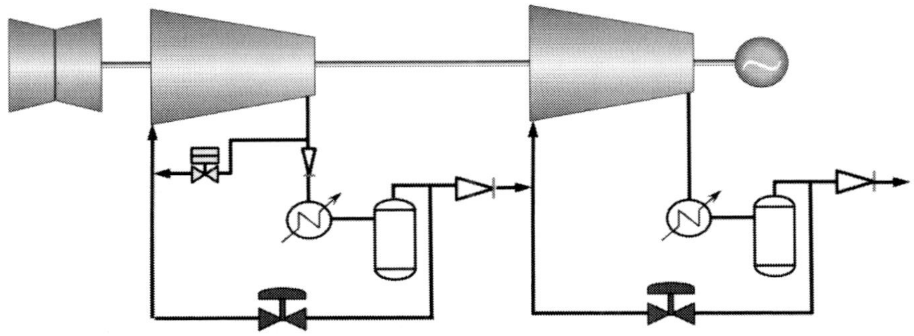

More Complex Machine Configurations

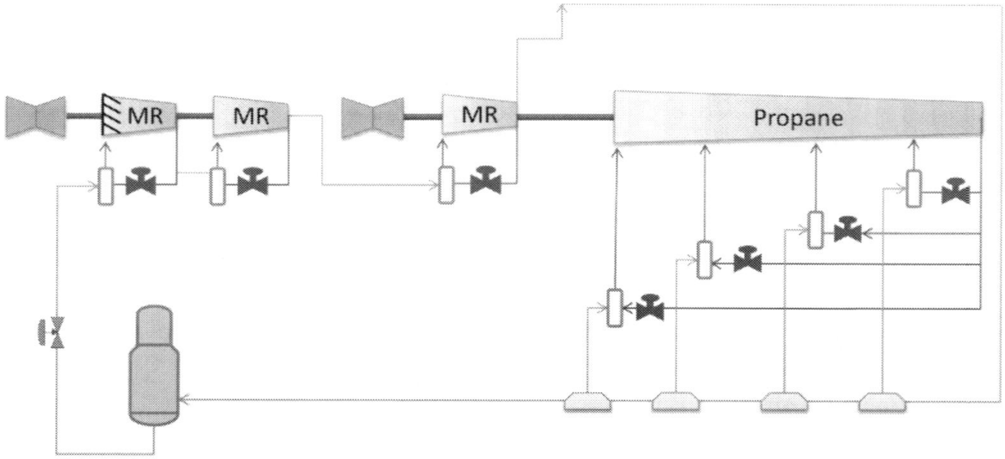

Split Shaft Compressor Design

- The Mixed Refrigeration LP and MP stages on one shaft while the HP stage on the Propane compressor shaft
  - Creates high dependency between the two compressor strings
  - A cascading trip can happen within seconds
- Typical cascade trip causes
  - Compressor surge trip
  - Underspeed trip due to high power limiting
- A requirement to keep either string online when the other trips
  - Avoid surging the compressor
  - Avoid excessive recycle that can overload the drivers

17 AIChE Spring Meeting, San Antonio, TX — 13TH GLOBAL CONGRESS ON PROCESS SAFETY

Compressor Control System Emulation

- Emulate the actual control system to ensure design and determine settings
- Controller emulator integrated with process and compressor train simulation
- Unload the running compressor when the other trips
  - Unload discrete input signal
  - Configurable Unload ramp rate and output clamp
  - Configurable hold timer
  - Allow Antisurge Control to override Unload sequence

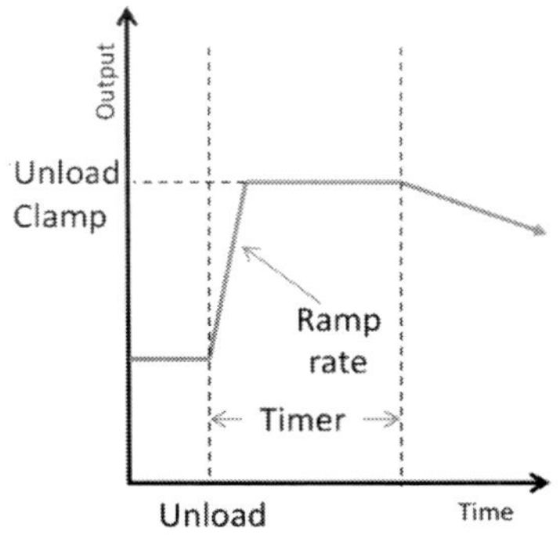

17 AIChE Spring Meeting, San Antonio, TX — 13TH GLOBAL CONGRESS ON PROCESS SAFETY

Simulate Tie In and Tie Out Procedures

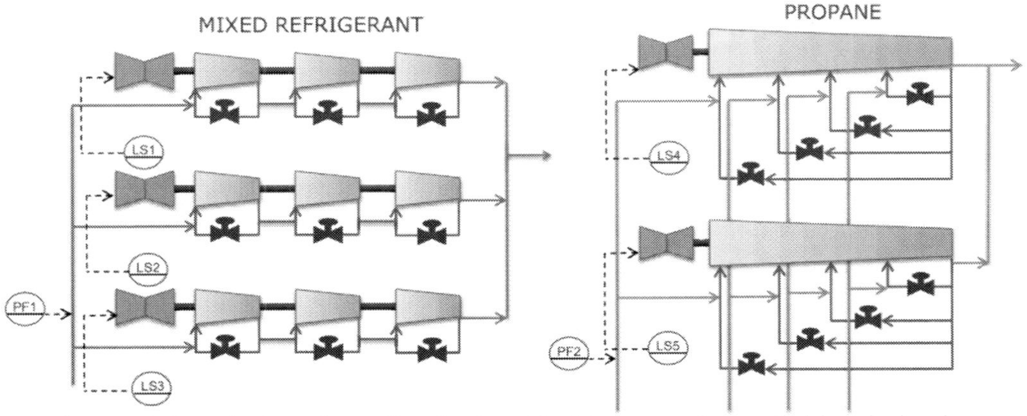

- Accurately mimic automatic startup/shutdown and parallel load-sharing control
- Verify operational procedures for bringing compressors on and offline

More Complex Machine Configurations

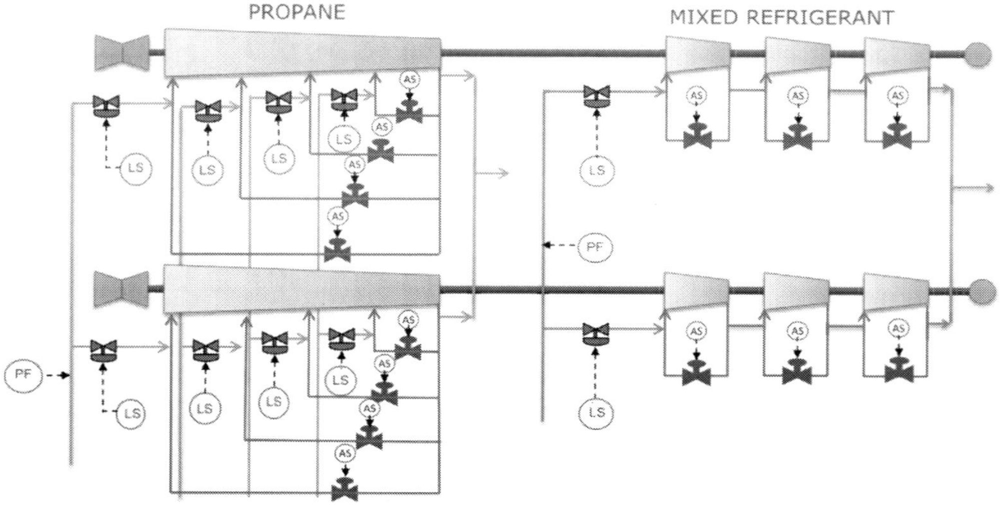

Complex Turbomachinery Control

- MR Compressor
  - Three Antisurge controls, one for each stage: LP, MP and HP
  - One Quench Control for LP
  - One Master Pressure Control for suction pressure Header
  - Loadsharing Control for each Compressor Train controlling the Inlet Throttle Valve
- PR Compressor
  - Four Antisurge controls, one for each stage: LP, MP, HP & HHP
  - Four Quench Controls, one for each stage: LP, MP, HP & HHP
  - Loadsharing Control for one Compressor Train; one for each inlet header: LP, MP, HP & HHP

17AIChE Spring Meeting, San Antonio, TX | 13TH GLOBAL CONGRESS ON PROCESS SAFETY

Optimize Process Control with Turbomachinery control

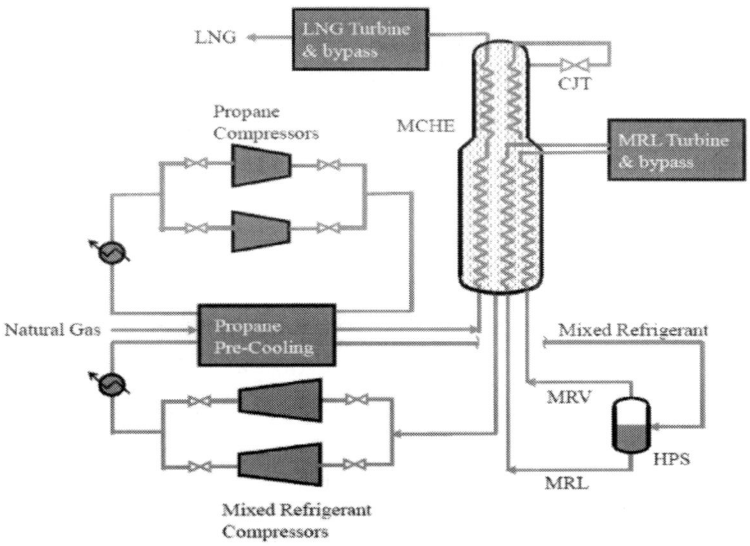

17AIChE Spring Meeting, San Antonio, TX | 13TH GLOBAL CONGRESS ON PROCESS SAFETY

Emulator Benefits

- Guarantees exact replication of the control system functionality
- Field engineers will not start tuning the system from default tuning values but will start from a model-tuned system.
- The configuration and tuning parameters can be easily transfered from real controllers to Emulator and vice versa.
- Guarantees more extensive tuning on estreme scenarios which are sometimes not tested in the field.
- The operator to get confident with the control system software tools (HMI, Configuration, or OTS use).
- Operating procedures and machinery sequencing (i.e. startup, load, unload, normal stop, etc.) can be defined or tested.

17 AIChE Spring Meeting, San Antonio, TX

13TH GLOBAL CONGRESS ON PROCESS SAFETY

Emulator System Overview

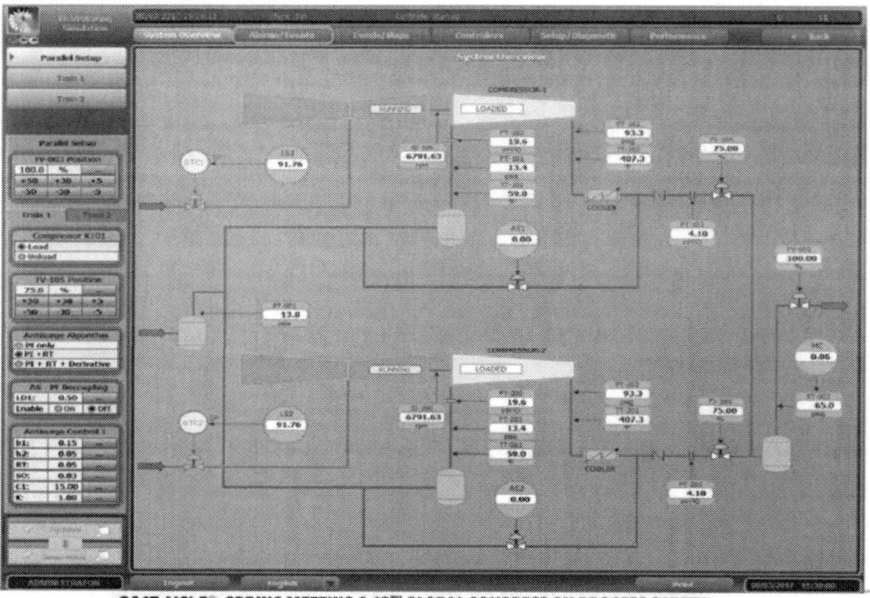

2017 AIChE® SPRING MEETING & 13TH GLOBAL CONGRESS ON PROCESS SAFETY

17 AIChE Spring Meeting, San Antonio, TX

13TH GLOBAL CONGRESS ON PROCESS SAFETY

THANK YOU

2017 AIChE® SPRING MEETING & 13TH GLOBAL CONGRESS ON PROCESS SAFETY

480254 Beyond the Rainbow: Practical Improvements in the Presentation and Communication of Hazard and Risk Analysis Data

Abstract:

Abstract

As technology improves and moves forward, so must the ways in which we communicate and present data. All too often, we as engineers, focus primarily on the numbers and neglect how our data is disseminated. Significant cost and effort is put forth to conduct engineering analyses, but the presentation and communication of the results to decision makers typically falls short. It is widely known that we (analysts) are capable of producing massive amounts of data and stuffing them into plot, graphs, and tables throughout hundreds of pages of reports. On the user end, how do we help key decision makers sift through this data, identify the critical parts, and better interpret and understand what we present?

The fundamentals of scientific visualization are primarily concerned with the ability to graphically illustrate scientific data to enable scientists to understand and glean insight into their data. With the continued rapid growth in computational power, the need to move beyond paper reports, tabular data, and rainbow scale contours plotted in two-dimensions is greater than ever. This paper will focus on reviewing the fundamentals and pitfalls of scientific visualization within quantitative hazard and risk analysis and provide practical examples for improvement. An examination of computational fluid dynamics (CFD) data, mixing of real images and calculated data, and QRA visualization of hazards in 3D with focus on an interactive user review will be presented. Each case will be discussed with real world examples, including a review of successes, failures, and where future improvements can be made.

Keywords: Simulation, QRA, Risk, modeling, hazard analysis

480331 Controlling Risk Ranking Variability By Using a Progressive Risk Registry

Abstract:

Controlling Risk Ranking Variability by Using a Progressive Risk Registry

Risk in the process industries is often overstated or understated due to insufficient information. This is especially true regarding Process Hazards Analyses (PHA) and incident investigations. Compliance audits often reveal risk ranking consistency issues even between identical hazards at the same operating facility. This session introduces the "Progressive Risk Registry" concept for controlling risk variability by integrating industry risk information and company-specific data with modeling.

This presentation will adress:

- What is a Progressive Risk Registry & how does it impact PHA?
- PHA Risk Ranking Variability
    - Case study
    - Unintended consequences
    - Management system failures
    - Lessons Learned
- Road to more accurate hazard consequences
    - Worst case credible events
    - Quantify extent of leak, fire, etc.
    - Risk Matrix controls
- Strategies for an Enterprise Level Progressive Risk Registry without the enterprise cost

Attendees will leave with new knowledge of what works in risk management along with real-world applicable tools and ideas to implement at their facilities.

- A common risk framework across the company?
- Ability to react to slow and fast changes?
    - MOC is key
    - Identify interdependencies and address them
- An understanding that mitigating one risk may adversely affect another?
- A way to effectively deal with unknown unknowns vs known unknowns (and a way to surface them)?
- The right people making risk decisions?
    - Make sure accountability is clear on who owns the risk
    - Focus on high outcome probability risks

480269 An Intelligent Supporting System for Safety Instrumented Systems Design for Enhanced Reliability and Availability

Abstract:

Safety Instrumented System design and operation is a complicated task and usually requires experts to perform it. Designing and successfully maintaining each phase of the Safety Life Cycle (SLC), defined by international standards, can also be time-consuming and very expensive. To avoid this problem and help the user to carry out analysis of minimal parts of the SLC, different supporting tools have been developed. However, these tools are complicated for the user and for some case, expensive. In this research, we propose and develop a SIS supporting tool for the analysis of the reliability of a given system, in the form of Android application. Our model avoids all the unnecessary user input parameters that have been challenging for the user. This reduction of parameters makes our tool easier to use, and compared to already developed tools the time required to perform reliability assessment becomes shorter. The supporting tool also contains a huge database of failure rates of Sensor, Logic Solver, and Final Element, included and continuously updated from different sources. This database can be modified or updated based on the user application area.

476834 Fundamental Insights in Industrial Chemical & Engineering Using Modeling & Simulation

Abstract:

Modern industrial chemical activities, i.e. discovery, development, management, design, scale-up, and manufacture, depend on computational technologies. Critical examples include process modeling and simulation, optimization of operations, and process control. More and more, the information supporting these tools comes from fundamental computations, e.g. computational fluid dynamics and molecular simulation. Modeling is useful because it complements data that cannot be obtained by experiment directly or easily. Moreover, many companies have successful efforts to accelerate the development of new and/ or better materials using an integrated blend of molecular modeling, simulation, and experimental techniques. This powerful scientific combination is proven capable of accurately predicting the physical, material, and performance properties of both established and new materials and processes.
This talk discusses how chemistry, micro-kinetic, and computational fluid dynamics modeling and simulation assisted the design, pilot, and production phases of a process and apparatus that removes tar from syngas of a breakthrough gasification and fermentation technology for conversion of biomass waste. This work lead to U.S. Patent 8,894,885 and a commercial-scale plant having an annual output of eight million gallons (24kta) of cellulosic ethanol and six megawatts of renewable power (Indian River BioEnergy Center). Key points are highlighted using examples from the chemical industry and the promise of industrial data analytics as a technical business tool is addressed.

479861 Decision-Making and Optimization of Steam Crackers Using COILSIM1D

Abstract:

Decision-Making and Optimization of Steam Crackers using COILSIM1D

process is carried out, it is also highly energy-demanding, amounting to approx. 8% of the entire chemical industry energy consumption. In this context, continuous adjustments to the daily operation are necessary in order to maximize production and margins, while accounting for variable feed types and qualities, market shifts and the usual operational limitations that influence the optimal operation point of each plant.

Steam cracking furnaces can be divided into two distinctive parts: radiant box and convection section. In the former, the process feed is cracked in tubular reactors suspended in a gas fired chamber whereas in the latter, the remaining energy content of the hot flue gas that results from the fuel combustion is recovered by preheating the process feed and other utility streams. As is to be expected, there is a very strong interlinking between radiant and convection section, hence influencing each other's operation. Determining settings for operation of each section that optimize the global process is, therefore, of utmost importance.

This paper focuses on the importance of furnace simulations that explicitly account for a detailed description of the convection section of steam crackers (see Figure 1), combined with dedicated heat transfer and evaporation models, as well as accurate physical property estimation methods. When integrated within a whole-furnace simulator, this level of detail allows to accurately represent the process, enabling to determine the optimal operation point that maximizes the combined efficiency of radiant and convection section operations, which is crucial in order to guide decisions oriented at optimizing the operation of furnaces.

To illustrate this, a case study of an ethylene production facility is presented, in which the operation of radiant and convection sections of the furnaces are optimized simultaneously. This task is straightforwardly carried out using the steam cracking simulation package COILSIM1D. This software counts with a detailed microkinetic model, combined with detailed convection section, furnace, TLE and steam drum models, enabling to maximize the yield of desired products while minimizing fuel consumption. This has largely positive economic and environmental implications in the process.

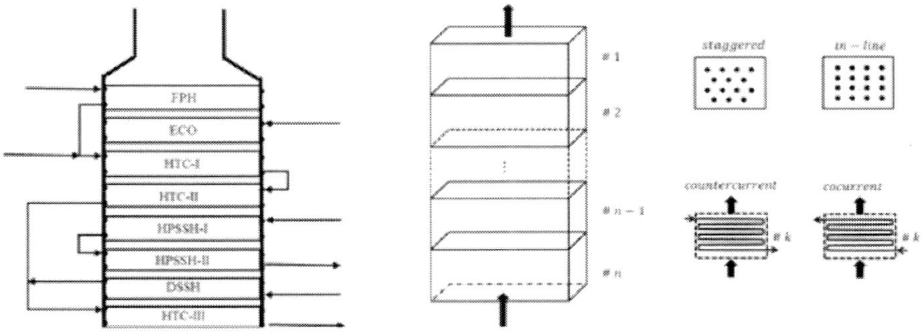

Figure 1: Schematic representation of the convection section of a steam cracking furnace. Various tube arrangements can be implemented.

AIChE Proceeding, Abstract #479869: 2017 Spring Meeting and 13th Global Congress on Process Safety, March 26-30, San Antonio, TX.

Lower Torque and Higher Efficiency Mixing with Contra-Rotating, Baffle-Free Impellers

Mr. Pongsarun Satjaritanun, Dr. Sirivatch Shimpalee, Dr. John W. Weidner, Dr. John R. Regalbuto, Department of Chemical Engineering, University of South Carolina, Columbia, SC

Mr. Eric Bringley, Department of Chemical Engineering and Biotechnology, University of Cambridge, Cambridge, United Kingdom

Dr. John A. Regalbuto, Perfect Mixing LLC, Columbia, SC

Abstract

This work investigates the intersection of two fields of work: (1) single axis, contra-rotating impellers and (2) buoyancy of solid suspensions. The objectives of this study are to (1) create a working model to quantitatively understand particle mixing, (2) characterize and compare contra-rotating single shaft impellers to single shaft dual impellers, and (3) improve quantification of particle mixing through image processing for both computational and experimental techniques. Twelve cases were studied by changing the direction of impeller rotation, impeller pumping direction, and the presence of baffles. Particles with Specific gravities (SG) of 0.866 and 1.05 were introduced into the experimental and computational systems in a finite and countable number. The numerical solutions were obtained by using the Lattice Boltzmann Method and the Discrete Particle Method. The evaluation of the input torques and particle dispersions versus rotating speeds were reported. Finally, the comparison of mixing efficiency were analyzed and concluded. This work will help understand mixing efficiency and provide justification for design decisions to improve industrial processes.

Introduction

Mixing and agitation operations are widely used in industrial process such as, pharmaceutical, agriculture, and food processing. Mixing is the manipulation of a heterogeneous physical system with the intent to make it more homogeneous [1, 2]. Mixing occurs when materials are agitated inside the reactor by several impellers. The contra rotating impeller was proposed for perfect mixing and energy saving. The contra rotating impeller consists of two propellers rotating in opposite directions about the same axis. Two propellers are arranged one behind the other, and the power is transferred from the motor via bevel gear, planetary gear or spur gear transmission [3]. Due to this mechanical structure, the contra rotating impeller has the advantages of high static pressure, high turbulent flow, and good performance in reversing fluid flow [4-6].

The twelve cases that were studied are shown in Figure 1. The mixing models were separated by the rotation patterns and reactor formats. Cases 1 to 4 are the contra rotating impellers without baffles. Cases 5 to 8 are the co-rotating impeller without baffles. Cases 9 to 12 are the co-rotating impeller with baffles in the tank. The tank geometry has a capacity of 25 Liters with a round bottom. The two propellers have a 4-pitched-blades at 45° and are driven by an adjustable rotation speed motor. These studies analyzed the motion of the solid particles in the liquid to find the solid-liquid mixing efficiency and the power consumption by using three-dimensional (3D) Computational Fluid Dynamics (CFD) modeling with the Lattice Boltzmann Method (LBM). The free surface model and Discrete Phase Model (DPM) was also used in this work. The solid-liquid mixing efficiency of the simulation can be calculated from the numerical results of the DPM by image analysis techniques [7]. Furthermore, the factors that impact the mixing of substance inside the reactor and their influences are discussed. The relationship between rotation configuration, rotation speeds and reactor designs that affects the power consumption to agitation and mixing efficiency are also reported.

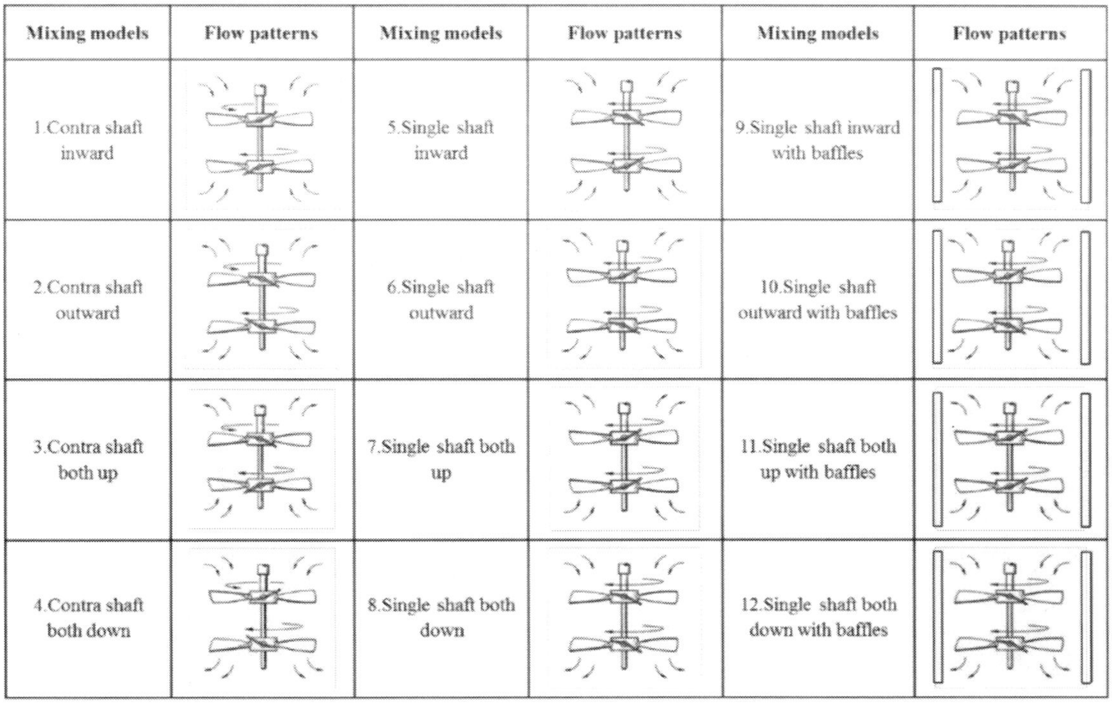

Figure 1. The twelve configurations used in study: Cases 1 to 4 contra rotating shaft, Cases 5 to 8 single rotating shaft, and Cases 9 to 12 single rotating shaft with baffles reactor.

Experimental setup and procedure

The experimental setup for this work is shown in Figure 2. It was constructed from two tanks, a cylindrical tank with a round bottom and a rectangular tank. The dimension of the cylindrical tank are 0.30 m in diameter and 0.35 m height. The dimension of the rectangular tank are 0.40 m in diameter and 0.40 m height. The cylindrical tank was installed inside the rectangular tank and filled with water because the rectangular tank can reduce the refraction and the illusion caused by the curvature of the cylinder tank which could affect the images. Four blade, 45° pitched blade turbine propellers were used and installed 10 cm. apart on the same shaft. The impeller was installed in the center of the tank and located 10 cm. above the bottom of tank. The impeller was stirred by a stirring machine with speed controls and a torque sensor. Two types of particles were used in this study at higher and lower density than water. The higher specific gravity had a diameter of 2.05 mm and specific gravity of 1.050. The lower specific gravity had a diameter of 2.45 mm and specific gravity of 0.866. The particles used in this study in each category contained 500 particles.

The particles movement in the tank was captured by two cameras on the front and side planes. When the system starts to rotate, a photo will be taken every 0.50 seconds from 0 to 40 seconds. Upon completion of the experiment, the prediction of particles inside the tank can be analyzed for mixing efficiency.

Example images from one of the cameras at the same point in time for different configuration are shown in Figure 3.

Figure 2. Experimental Apparatus used to evaluate torque and capture experimental data.

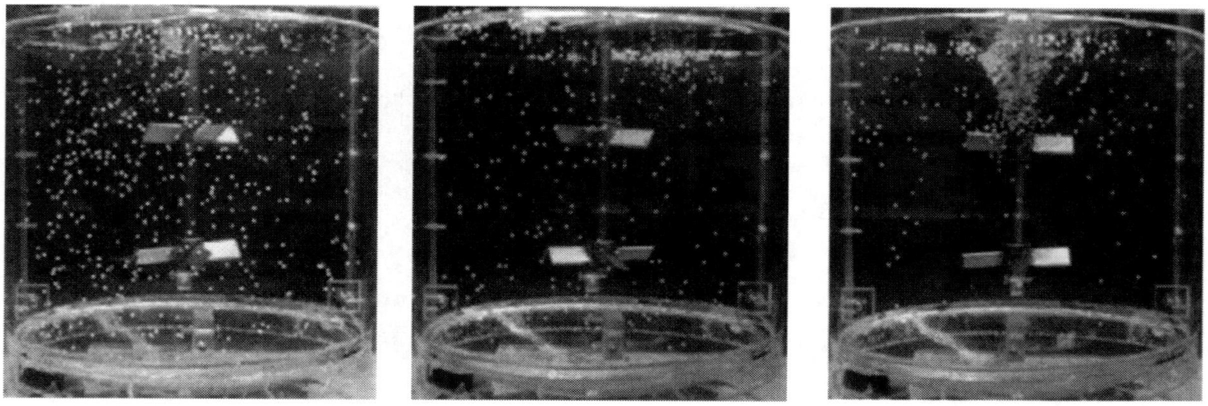

(a) (b) (c)

Figure 3. Particles inside the stirring tank under different dispersions.

Image analysis

Tracking particles as they are drawn down from the liquid surface through image analysis provides an easy way to calculate the particle concentration and mixing efficiency in transparent systems. This study presents a quantification scheme for countable systems and the calculation takes into account the homogeneity of the system's concentration through the locations of the particles. Images were taken of cross sectional planes, as seen in Figure 3, where there is high contrast between the background and the particles submerged.

These 2D images are divided into horizontal and vertical regions to measure the dispersion of the particles depth and spread respectively. The regions are assigned a weight according to the volume they represent and the anticipated fraction of visible particles for a perfectly homogeneous system. A region that represents 25% of the fluid volume is anticipated to have 25% of the particles when homogeneously mixed. The systems are assumed to be dilute enough such that overlap is not significant. An example can be seen in Figure 4 with division and weights.

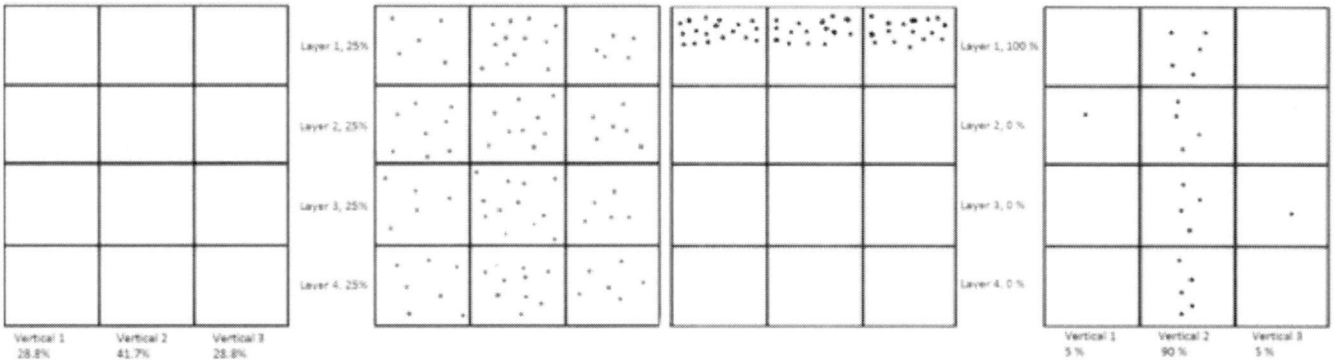

Figure 4. Cross section division with examples from left to right of the assigned weights, good mixing (unquantified), poor drawdown, and poor dispersion. The two right hand images would generate strong deviation terms making their degree of deviation close to zero because of the poor mixing.

After image processing, the number of particles visibly drawn into the liquid (N_{vis}) for the horizontal and vertical regions are recorded. The deviation from ideal mixing (d_i), is calculated individually for each horizontal (h) and vertical (v) regions; the square root of the sum of square differences between the ideal and actual particle fraction (f) as shown in Equation (1). The calculation of the particle fraction from those images accounts for the area of solid particles per the area of fluid. The worst case d_i^* is taken when all of the particles are in the smallest region, thus maximizing the deviation from ideal.

$$d_i = \sqrt{\sum (f_{ideal} - f_{act})^2}\,, i = h, v \tag{1}$$

The degree of deviation (δ_i) is calculated as one minus half the ratio of d_i and d_i^* shown in Equation (2). Because of the maximized d_i^*, the range of this equation will be between 0.5 (poor mixing) and 1 (near-perfect mixing).

$$\delta_i = 1 - 0.5\left(\frac{d_i}{d_i^*}\right)^2, i = h, v \tag{2}$$

The mixing efficiency (μ) is defined in Equation (3) as the fraction of particles visible and the average product of the degree of deviation in the horizontal and vertical direction over the two planes. Because the range of δ_i is between 0.5 and 1, the greatest penalty is a 75% reduction. A quadratic function was chosen to keep a low gradient for slight deviations and high for large deviations from homogeneous, making the combined penalty non-linear.

$$\mu = \frac{N_{vis}}{N_{act}} \cdot \frac{(\delta_{v,1} \cdot \delta_{h,1} + \delta_{v,2} \cdot \delta_{h,2})}{2} \tag{3}$$

Model development

Lattice Boltzmann method (LBM)

LBM was chosen to perform the numerical analysis in this work. It is one of the most powerful techniques for computational fluid dynamics (CFD) for a wide variety of complex turbulent flow problems including multiphase flow and free surface models in complex geometries. This method uses the concept of streaming and collisions of particles that incorporate the physics of microscopic and mesoscopic processes so that the macroscopic averaged properties obey the desired macroscopic equations [8-15]. LBM schemes are classified as a function of the spatial dimensions m and the number of distribution functions n, resulting in the notation *DmQn*, respectively. For the three dimensional (3D) model used in this work, a commercial LBM solver, XFlow 2016 (Build 98), was chosen to perform the calculation. This solver uses spatial dimension of 3 and number of distribution functions of 27 (i.e., *D3Q27*).

Computational model

3D time-dependent simulations of the mixing and agitation system with the free surface model were ran for the twelve different designs: Cases 1 to 4 (Contra-rotating impeller without baffle tank), Cases 5 to 8 (Co-rotating impeller without baffles tank), and Cases 9 to 12 (Co-rotating impeller with baffles tank) as shown in Figure 1. The standard geometry involves an impeller and tank. The impeller has two propellers and can rotate in the opposite or same direction. The tank has a cylindrical shape and the bottom of tank has a round shape. The system configuration is shown in Figure 5.

As mentioned earlier, the commercial software based on Lattice Boltzmann method was used in this study. The scale of lattice element was set at 2 mm and the time to study was arranged for 40 seconds. The simulation settings are shown in Tables 1.

In this study, the particles have properties as follow: sphere shape, specific gravity of 0.866 and 1.050, particle diameter of 2.45 mm and 2.05 mm, respectively, and 500 particles. The injector surface is on the bottom of the tank for high specific gravity and on the top liquid surface for low specific gravity. The system start feeding the particles when the simulation starts.

Table 1. The simulation specifications of the stirring model.

Variables	Value	Unit
Rotation speeds, n	150, 200, 250, 300, 400	rpm
Water temperature	20	°C
Water density, ρ	998.3	kg/m^3
Water dynamic viscosity, ν	1×10^{-3}	Pa.s
Water diffusion coefficient	2.26×10^{-9}	m^2/s
Water volume	21.204	liters
Solid particle density, ρ	866 and 1050	kg/m^3
Solid particles number	500	-
Solid particle size	2.500 and 2.050	mm
Gravity, g	9.810	m/s^2

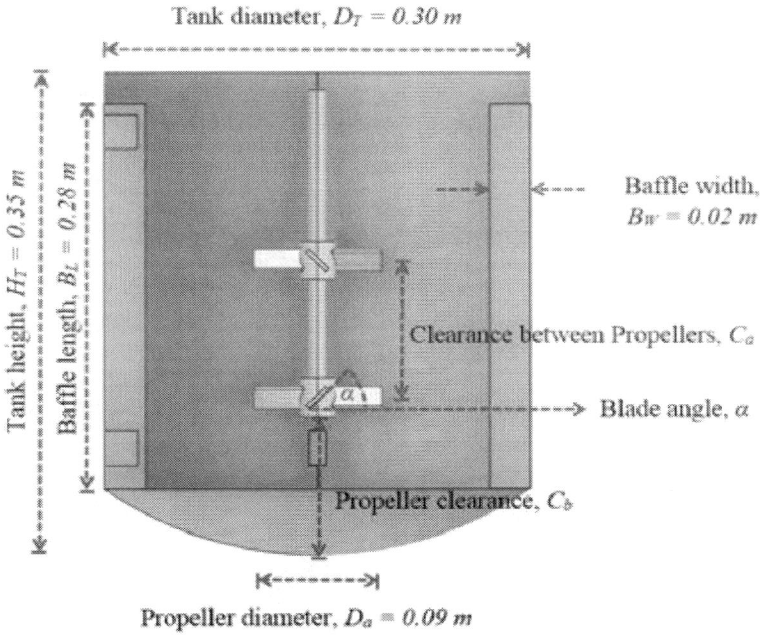

Figure 5. The representation of reactor/impeller dimension for the varying twelve cases.

Result and discussion

Figures 6a-d present the stirring torque validation between simulation results and experimental data for the twelve cases. In the experiment, we used a stirring machine with a torque sensor to evaluate the power consumptions. In the simulation, we calculated the force on the blades which the resulting momentum allowed us to evaluate the torque. In the single co-rotating impeller without baffle (Cases 5 to 8), the results show these systems require lower power. This could be due the result of the vortex and the bulk flow in the reactor. When baffles are installed in the reactor (Cases 9 to 12), these baffles eliminate the vortex meanwhile increasing the required torque [7]. This system requires more power than the system with single co-rotating impeller without baffle.

Both the results of the single co-rotating impeller with and without baffle can explain the relationship between torque and flow behavior in the reactor. The results of contra-rotating impeller without baffle (Cases 1 to 4) show these systems require higher torque than other cases, which is caused by the high turbulent flow in the reactor. All of the cases in this study, the stirring torque increases with increasing rotation speed. This is because of the effect of high turbulent fluid flow in the reactor. The torque required for each system does not have a direct relation to mixing efficiency. Due to its design significance, mixing efficiency should be studied as a function of torque for each geometry.

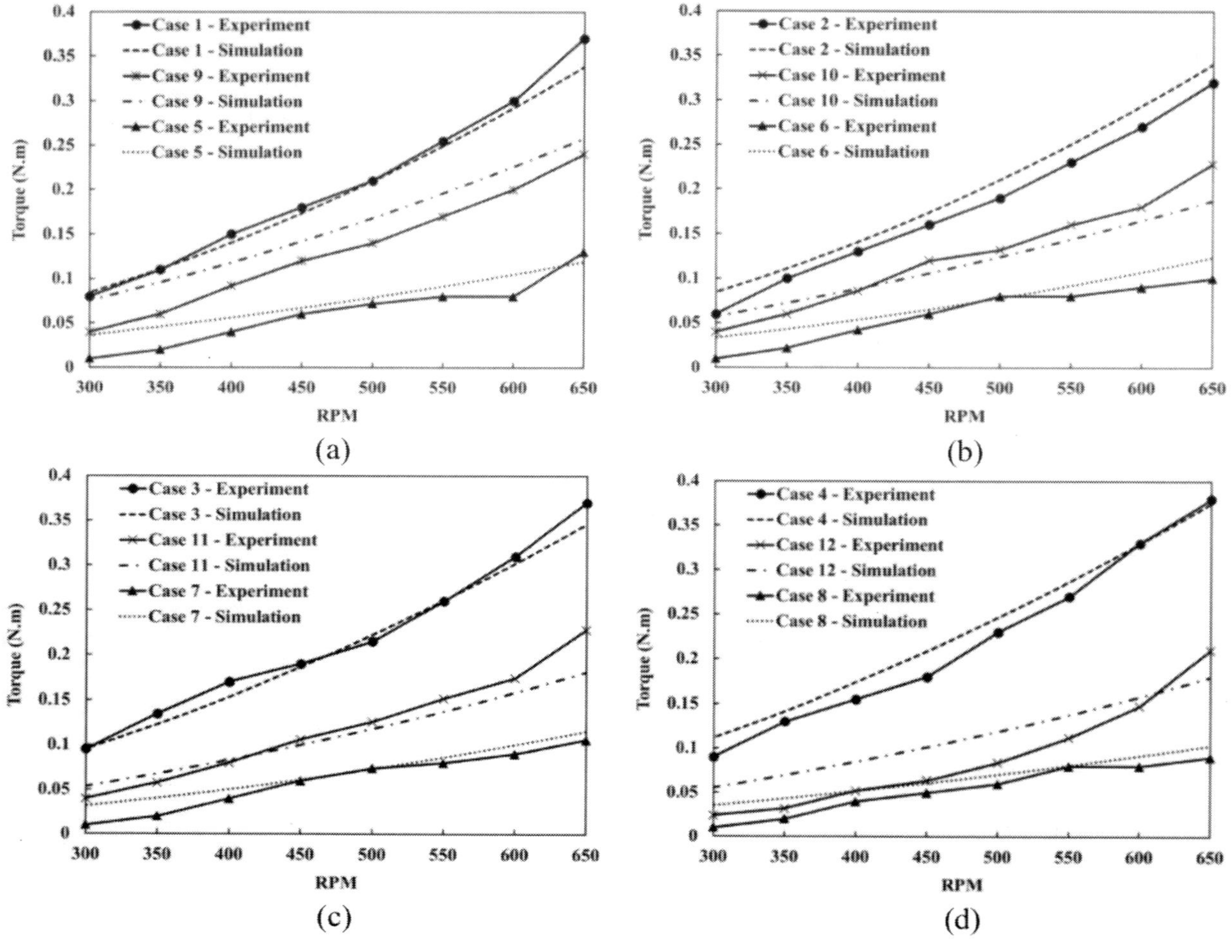

Figure 6. Simulation and experimental stirring torque compared in different reactor designs at rotation speeds from 300 to 650 rpm (a) Inward impeller configuration (b) Outward impeller configuration (c) Both-up impeller configuration (d) Both-down impeller configuration.

The visual observation presents the qualities of particles dispersed but not the quantities. Therefore, the image analysis technique method was used to evaluate the mixing efficiency as explain above. This method can provide the mixing efficiency of both experimental raw data and simulation as shown in Figure 7. The mixing efficiency can be indicated by the particle movement and uniform dispersion throughout the tank. Figures 7a-f present the transient simulation and experimental mixing efficiency of Cases 1, 5 and 9 at different specific gravity. The rotational speed was set at 300 rpm for all models. These models have the same flow configuration (i.e., inward flow direction). For the SG 0.866, the mixing efficiency of simulations are higher than experimental data at the time during 1 to 20 seconds. This is due to the effect of different particles seeding area on the top of liquid surface in the simulations.

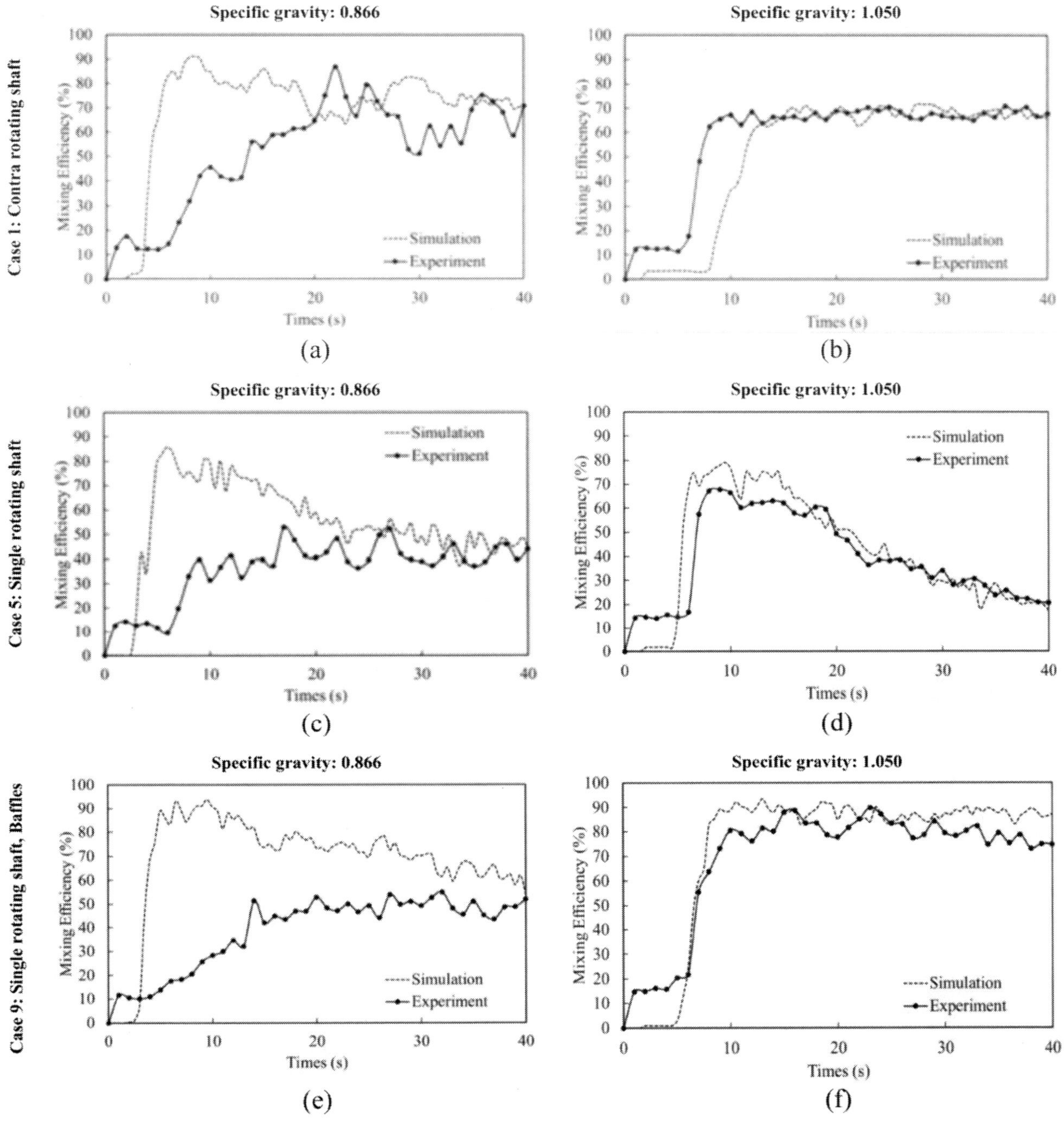

Figure 7. Simulation and experimental mixing efficiencies compared in different reactor designs at two different specific gravities at a rotation speed of 300 rpm.

Figures 8a-b compares the simulation and experimental data at steady state mixing efficiency for all design at the rotation speed of 300 rpm. Figure 8a presents the mixing efficiency at SG 0.866. The result shows similar mixing efficiency for both the experimental and simulation results. As well as, Figure 8b presents the similar mixing efficiency at SG 1.050 for all cases. The cases of co-rotating impeller (Cases 5 to 8) show the low mixing efficiency at both specific gravities. Because these tanks do not have any baffle inside the tank thus causing the formation of vortex that producing only the bulk flow in the tank. The contra rotating impeller cases (Cases 1 to 4) and co-rotating impeller with baffles tank cases (Cases 9 to 12) show the better mixing efficiency than Cases 5 to 8. Because these designs have more turbulent flow in the tank. However, the co-rotating impeller with baffles tank gives more mixing efficiency than the contra rotating at higher specific gravity. Therefore, the contra rotating impeller is suitable for the low specific gravity and the co-rotating impeller with baffles tank is appropriate for high specific gravity.

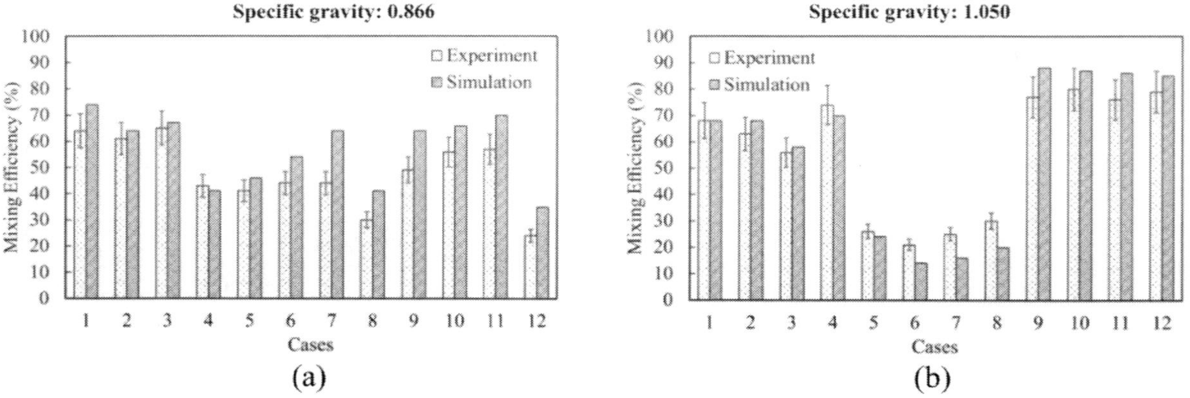

Figure 8. Simulation and experimental mixing efficiencies compared in different twelve reactor designs at two different specific gravities at a rotation speed of 300 rpm.

Conclusion

Contra rotating impellers system requires extra energy, but they will have better mixing of low specific gravity particles. Power consumption increases with increasing rotation speed. This is because as the high speed turbulent fluid velocity fluctuates inside the tank, it hits the propeller blades which makes an inconsistent force against the impeller.

Mixing efficiency can be calculated by numerical solutions by Lattice Boltzmann method combining with the image analysis technique. This technique can be used to better understand mixing characteristics in a wide variety of industrial applications. From the results above, the rotation speed and flow configuration play major role on mixing characteristic, particularly when buoyancy forces are significant. The bulk motion of a vortex causes poor mass transfer and poor mixing. However, the vortex can be eliminated by using contra rotating impeller or the installation of baffles around the tank. Moreover, the contra rotating impeller can eliminate the dead zone at the groove corner of baffles. From the foregoing, the variable that is important to the design selection is power consumption and mixing efficiency. Further study on the relationship between time averaged flow profiles and mixing characteristics are recommended.

Acknowledgment

The authors would like to thank Perfect Mixing LLC for providing the experimental equipment. The authors also would like to acknowledge Next Limit Technologies Company for providing the XFlow software.

References

1. Dantec Dynamics A/S, https://www.dantecdynamics.com/mixing.
2. E.L. Paul, V. Atiemo-Obeng, S.M. Kresta, Handbook of Industrial Mixing: Science and Practice, Wiley (2003).
3. A.F. El-Sayed, Fundamentals of Aircraft and Rocket Propulsion, Springer (2016).
4. Q. Wang, D. Meng, A study of the matching of impellers and motors for contra-rotating fan based on electromagnetic-fluid coupling analysis, Advances in Mechanical Engineering 8 (2016) 1-11.
5. K.J. Paik, S. Hwang, J. Jung, T Lee, Y.Y. Lee, H. Ahn, S.H. Van, Investigation on the wake evolution of contra-rotating propeller using RANS computation and SPIV measurement, International Journal of Naval Architecture and Ocean Engineering, 7 (2015) 595-609.
6. K.S. Min, B.J. Chang, H.W. Seo, Study on the Contra-Rotating Propeller system design and full-scale performance prediction method, International Journal of Naval Architecture and Ocean Engineering, 1 (2009) 29-38.
7. P. Satjaritanun, Y. Khunatorn, N. Vorayos, S. Shimpalee, E. Bringley, Numerical analysis of the mixing characteristic for napier grass in the continuous stirring tank reactor for biogas production, Biomass and Bioenergy 86 (2016) 53-64.
8 U. Frisch, B. Hasslacher, Y. Pomeau, Lattice-gas automata for the Navier-Stokes equation, Physical review letters, 56(14) (1986) 1505-1508.
9 G.R. McNamara, G. Zanetti, Use of the Boltzmann equation to simulate lattice-gas automata, Physical Review Letters, 61(20) (1988) 2332-2335.
10 S. Chen, GD. Doolen, Lattice Boltzmann method for fluid flows, Annu. Rev. Fluid Mech, 30(1) (1998) 329-364.
11 X. Shan, H. Chen, A general multiple-relaxation-time Boltzmann collision model, International Journal of Modern Physics C, 18(4) (2007) 635-643.
12 D. d'Humieres, Multiple relaxation time Lattice Boltzmann models in three dimensions, Philosophical Transactions of the Royal Society of London. Series A: Mathematical, Physical and Engineering Sciences, 360(1792) (2002) 437-451.
13 DM. Holman, R. Brionnaud, Z. Abiza, Solution to industry benchmark problems with the Lattice-Boltzmann code XFlow, Proceeding in the European Congress on Computational Methods in Applied Sciences and Engineering (ECCOMAS), Vienna, Austria, 2012.
14 K. Premnath, S. Banerjee, On the three-dimensional central moment Lattice Boltzmann method, Journal of Statistical Physics, 143(4) (2012) 747-761.
15 Next Limit Technologies, XFlow CFD, http://www.xflowcfd.com.

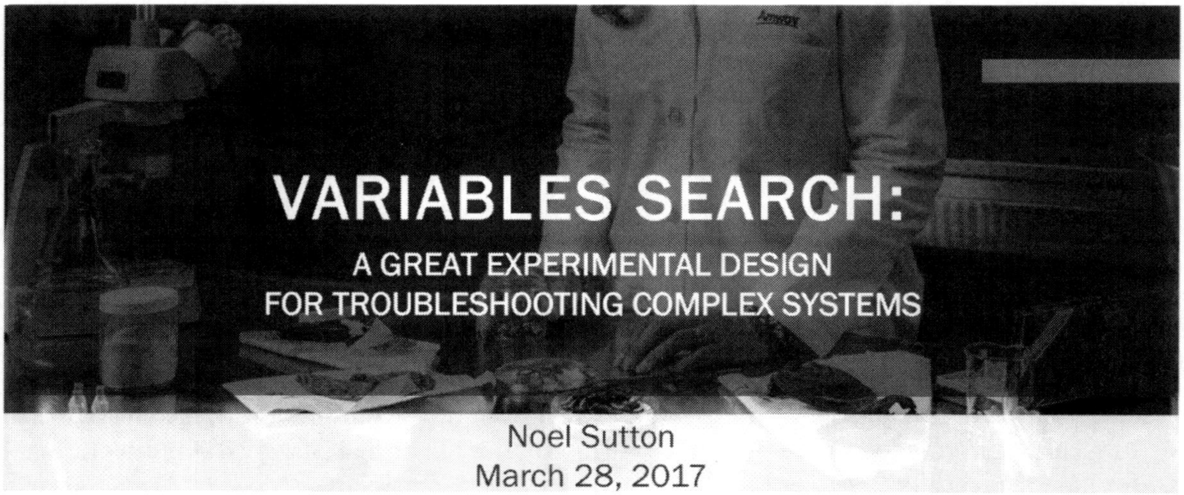

VARIABLES SEARCH:

A GREAT EXPERIMENTAL DESIGN FOR TROUBLESHOOTING COMPLEX SYSTEMS

Noel Sutton
March 28, 2017

Amway
RESEARCH & DEVELOPMENT

1

What is an Experimental Design or DoE?

Amway
RESEARCH & DEVELOPMENT

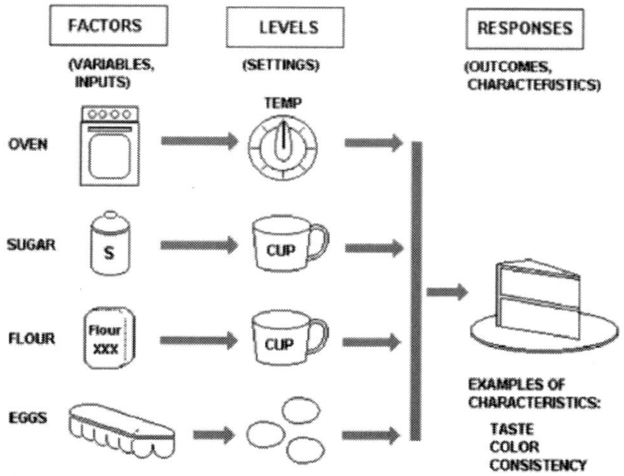

2

Full Factorial Experimental Design

Amway RESEARCH & DEVELOPMENT

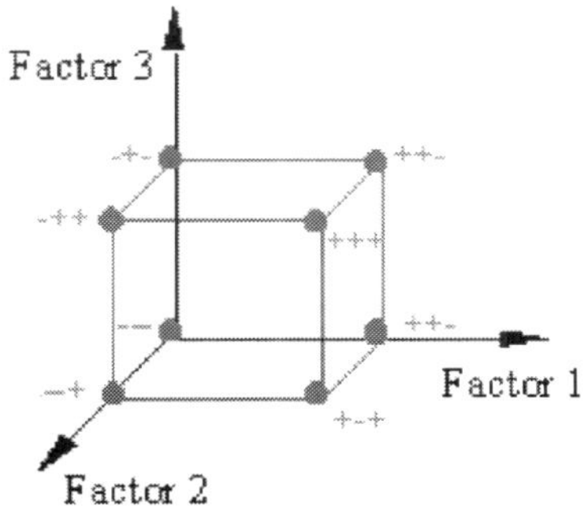

3

Why Use Variables Search?

Amway RESEARCH & DEVELOPMENT

Compared to other experimental designs, VS is:

- Much easier to learn & use
- For complex systems, relatively few experiments are required to pinpoint the critical variables
- Main & interaction effects are not "confounded"
- One quickly finds out whether key parameters have been included in the DoE or overlooked
- Some experiments are replicated – which shows whether results are repeatable, & enables one to gauge the magnitude of experimental error

4

Comparison of Experiments Required

Amway RESEARCH & DEVELOPMENT

Experimental Design Method	Number of Variables 4	5	6	7	8
2 LEVEL FULL FACTORIAL: no confounding, significant variables may not have been included in DoE, no replication	16	32	64	128	256
HALF-FRACTION FACTORIAL: some confounding, significant variables may not have been included in DoE, no replication **(Problems noted above also apply to other DoEs like PLACKETT-BURMAN or TAGUCHI ORTHOGONAL ARRAYS)**	8	16	32	64	128
VARIABLES SEARCH: no confounding, one determines very quickly whether significant factors have been included in or left out of the DoE, some experiments replicated	10-16	10-18	10-20	10-22	10-24

5

Variables Search Phases

Amway RESEARCH & DEVELOPMENT

- Phase 1 - Determine whether the critical factors have been included in the DoE
- Phase 2 – Expose the Critical Variables
- Phase 3 - Confirms that all key factors have been uncovered

6

Phase 1 Steps

Amway RESEARCH & DEVELOPMENT

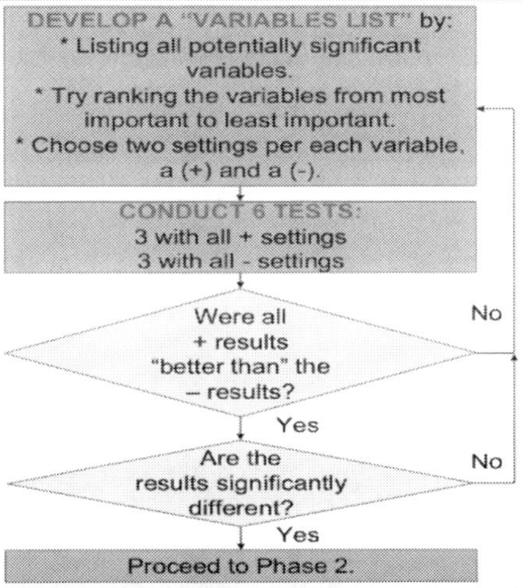

7

Variables List Example

Amway RESEARCH & DEVELOPMENT

	(+) Beneficial Conditions	(-) Detrimental Conditions
Turbine Agitator Speed during Cooling	Slow	Fast
Scrapewall Speed during Cooling	Slow	Fast
Drop Temperature	105-110'F	90-95'F
Shearing Finished Product (mill and/or versate)	No shear	High shear
Mixing Time After KOH Addition	15 min	45 min
Cooling Rate	Fast	Slow

8

Tips for creating a "Variables List":

- *Selecting Variables*: include, rather than exclude, seemingly irrelevant variables because those variables might emerge as significant.
- *Ranking Variables, from most to least important*: since the most important variables will be checked 1st during Phase 2, the prescribed number of experiments will be minimized if ordering is done well.
- *Choosing Settings*: each (+) value should be chosen with the belief it could lead to the desired outcome, while each (-) value should be selected with the thought it could lead to an unwanted result.

9

Phase 1 Steps

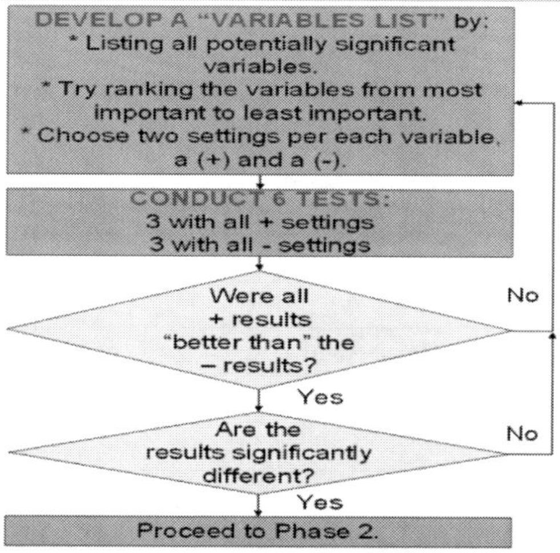

10

Phase 1 Experiments Example

Amway
RESEARCH & DEVELOPMENT

	Variables					
	A	B	C	D	E	F
Experiment 1	+	+	+	+	+	+
Experiment 2	-	-	-	-	-	-
Experiment 3	-	-	-	-	-	-
Experiment 4	+	+	+	+	+	+
Experiment 5	+	+	+	+	+	+
Experiment 6	-	-	-	-	-	-

11

Successful Phase 1 Results

Amway
RESEARCH & DEVELOPMENT

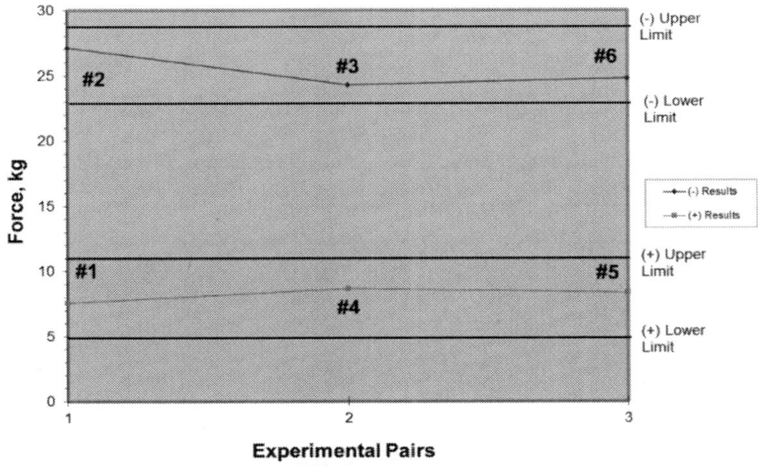

12

Phase 2 Steps

Amway
RESEARCH & DEVELOPMENT

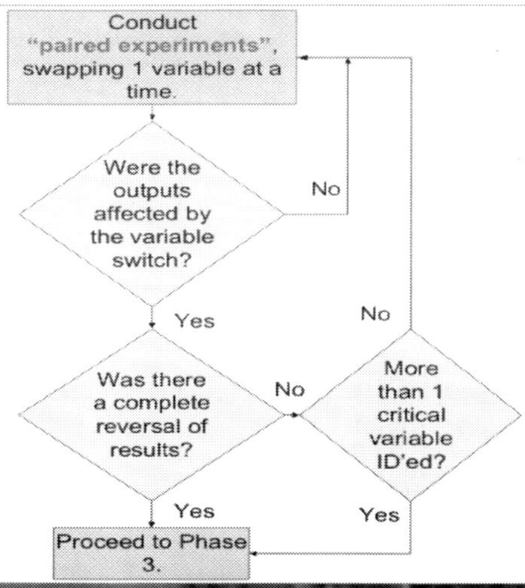

13

Paired Experiments Focusing On Variable A

Amway
RESEARCH & DEVELOPMENT

	Variables					
	A	B	C	D	E	F
Experiment 7	+	-	-	-	-	-
Experiment 8	-	+	+	+	+	+

14

Example of an Insignificant Variable

Amway
RESEARCH & DEVELOPMENT

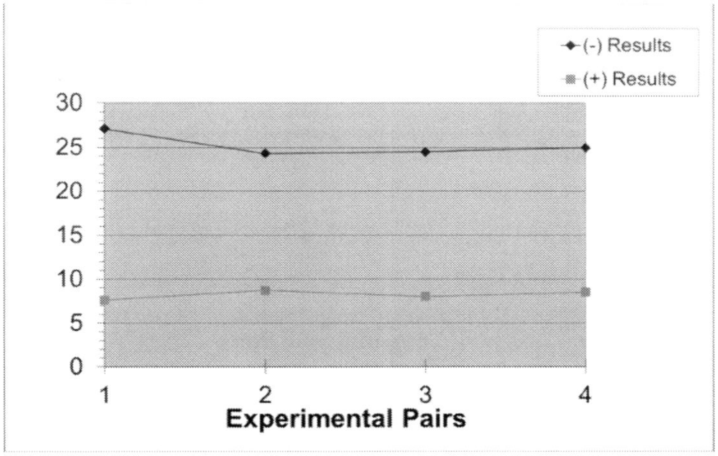

15

The Critical Variable is Uncovered

Amway
RESEARCH & DEVELOPMENT

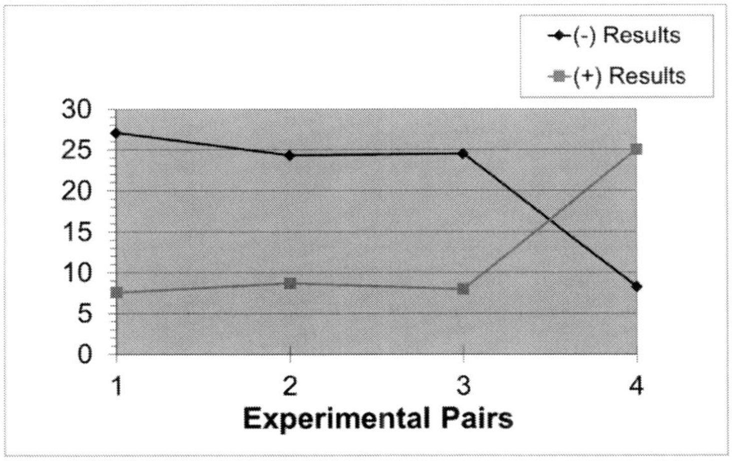

16

An Interacting Variable is Found

Amway
RESEARCH & DEVELOPMENT

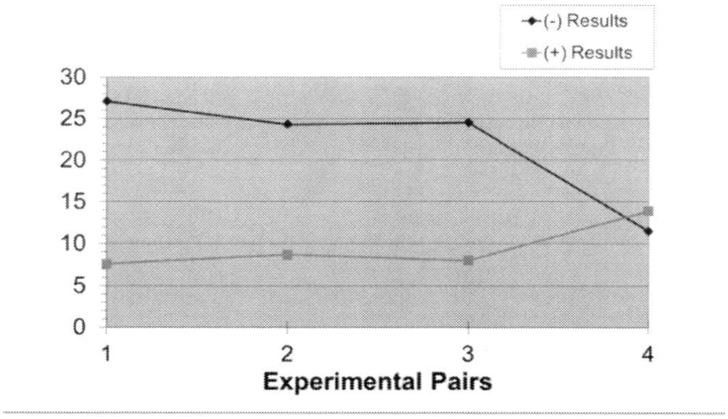

17

Phase 2 Steps

Amway
RESEARCH & DEVELOPMENT

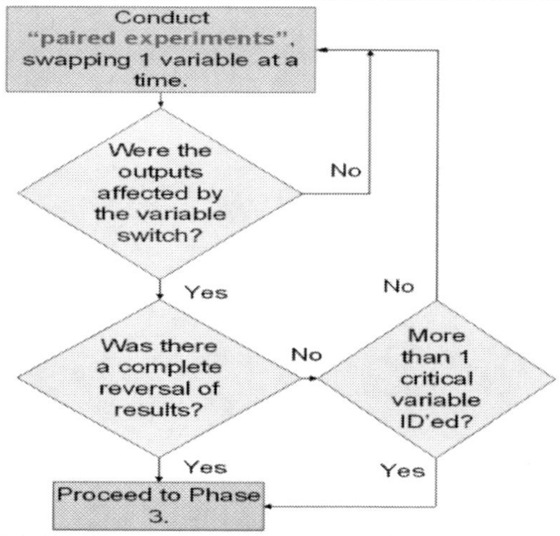

18

Phase 2 – 2nd Interacting Variable Identified

Amway
RESEARCH & DEVELOPMENT

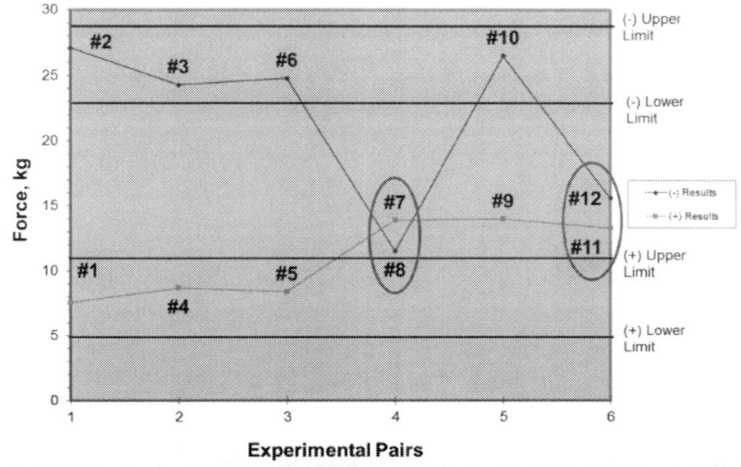

Phase 3 - Confirm All Key Factors have been ID'd

Amway
RESEARCH & DEVELOPMENT

	Variables						
	A	B	C	D	E	F	Response
Experiment 13	-	+	-	+	+	+	Same as (-) results in Phase 1
Experiment 14	+	-	+	-	-	-	Same as (+) results in Phase 1

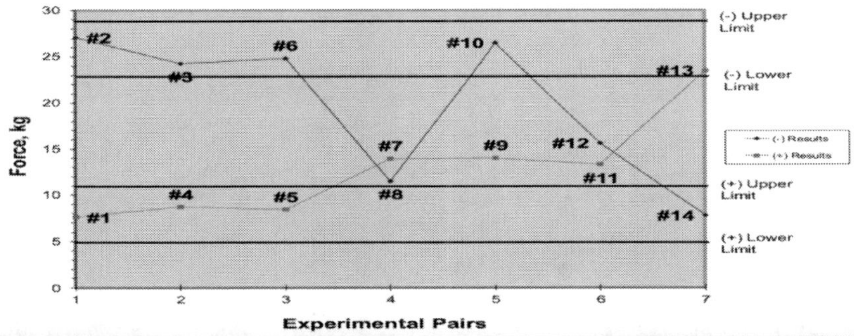

Quantifying & Controlling Significant Variables

Amway
RESEARCH & DEVELOPMENT

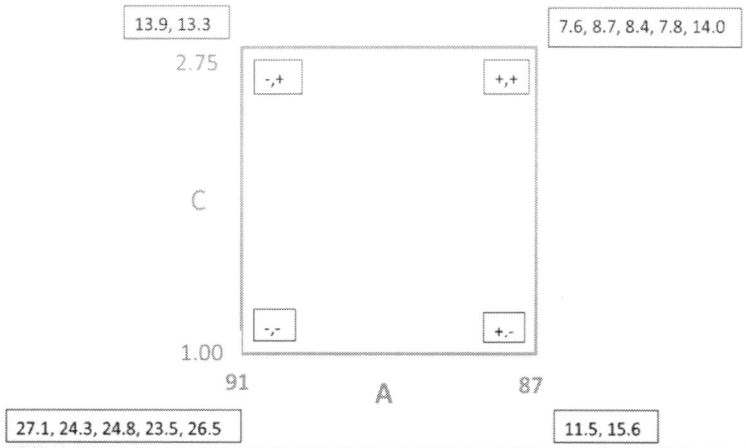

21

Viscosity Study

Amway
RESEARCH & DEVELOPMENT

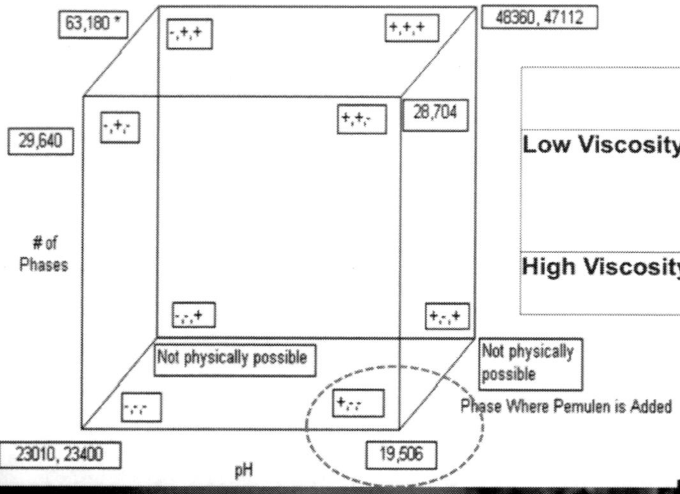

	pH	Phases	Pemulen Add'n
Low Viscosity (-)	7	One	Premix w/ Finsolv & then add to water phase
High Viscosity (+)	6	Two	Add to oil phase

22

Benefits of Variables Search

Amway
RESEARCH & DEVELOPMENT

- Easy to learn & use
- Relatively few experiments are required to pinpoint the critical variables in complex systems
- Main & interaction effects are not "confounded"
- One determines quickly whether key parameters have been accounted for or overlooked
- Some experiments are replicated – which reveals whether results are repeatable, & enables one to gauge experimental error

Noel Sutton
noel_sutton@amway.com
616-787-5714

479904 Validation and Application of Process Models for Solving Problems in Chemical Processes

Abstract:

Process models are extensively used for evaluating new plant designs and optimization of existing assets. In order to successfully use these models, they should be validated against pilot or industrial plant data. The validation of such models is not a trivial exercise due to their complexity and uncertainties in the data. A validation approach using equation oriented method will be discussed. The approach will be illustrated with examples from the author's practical experience. The examples cover application of the suggested methods to highly heat integrated distillation system and industrial closed loop refrigeration process.

480972 Single and Multi-Objective Optimizations Using Existing Rigorous Process Simulations: A New Hope

Abstract:

Today, traditional process simulation software does not take full advantage of advances in CPU architecture (multiple cores) for parallel processing. At the same time, industrial needs for achieving optimum process conditions for both single and multiple objectives are at their greatest. This led many industrial and academic groups to investigate equation-oriented (EO) solvers, custom written code, and/or reduced models to achieve the necessary calculation speed to perform large scale optimizations. The disadvantages have been extended setup time, significant domain expertise requirements, difficult model maintenance, and sub-optimal solutions.

This paper will discuss significant improvements in calculation time achieved through parallel processing techniques in a traditional sequential process simulation package combined with advanced optimization methods to ensure global optimums in single and multiple objective optimizations. The industrial benefit is a reduction in setup time, a reduction in required expertise, the ability to leverage existing process models, and the benefits of first principles' simulation. Calculation time is competitive with rigorous EO methods. We will present examples of single and multiple objective optimizations of industrial processes and discuss setup and calculation times.

480328 Startup Optimization and Flexibility Analysis for Ethylene Procedures

Abstract:

Startup procedures of ethylene plants normally take several days, consume million pounds of materials, and generate a lot of flare emissions. Startup procedures also face many uncertainties. A single facility failure may lengthen the whole startup and dramatically increase flare emissions.

In our previous works, startup procedures have been simulated by dynamic simulation, and several improvement procedures have been proposed to reduce the startup time and flare emissions. Based on previous works, we build a MIP (Mixed-Integer Programming) model to optimize the schedule of the whole startup procedure. In this model, the startup procedure is divided into many steps of facilities, which can be re-scheduled by the MIP model to get optimized startup time and flare emissions.

The MIP model can also handle flexibility issues during startup. Possible delays of key steps are listed in the model, and mathematical programming can re-schedule subsequent steps to keep overall startup time minimal.

A case study is also presented. Comparing to base case, the optimized case reduces startup time by 20%, and reduced overall flare emissions by 50%. Moreover, even with delays of several key facilities, the total startup time and flare emissions are still less than the base case.

AUTHOR INDEX